HARVARD MONOGRAPHS IN THE HISTORY OF SCIENCE

THE DISCOVERY OF THE
CONSERVATION OF ENERGY

The Discovery of the Conservation of Energy

YEHUDA ELKANA

With a Foreword by
I. BERNARD COHEN

HARVARD UNIVERSITY PRESS
Cambridge, Massachusetts

© 1974 by Yehuda Elkana

Second printing 1975

Library of Congress catalog number 73–88897
ISBN 0-674-21240-1

Printed in the United States of America

In the world of human thought generally and in physical science particularly, the most fruitful concepts are those to which it is impossible to attach a well-defined meaning

H. A. KRAMERS

To Yehudit, and our friends who helped me morally and otherwise when I was at the crossroads

Contents

Preface

The main idea of this book is the meeting of several converging lines of thought. The fundamental idea behind it all is the essential unity of all aspects of the intellectual enterprise. This idea, though it sounds like an overworked commonplace, can serve as a practical guide to historical investigation. Above all, it implies a coherence theory of personality. That same unidentifiable element which makes one recognize any piece by Bach, Beethoven or Britten is there in any written work too. Newton's *Principia*, Newton's alchemical studies and Newton's theological works are demonstrably written by the same man. Books come to illustrate this: to illustrate the regular interaction between science and philosophy, by a case history. The interaction, I assume, is an influence which works in both directions. 'The Emergence of the Concept of Energy' seems to me to be one chapter in a long unwritten history which illustrates the roots of science in metaphysics and the roots of scientific metaphysics in the problem-situations of the age: this book is that chapter.

Concept-creation originates in basic metaphysical principles, which cannot be empirically tested. This is not to say, of course, that a metaphysically orientated philosopher-scientist does not ever do experiments; rather, it means that he will interpret experiments, his or another's, in the light of his metaphysical principles, instead of testing his hypotheses by systematic 'conjectures and refutations'.

It was such a metaphysical belief, in the principle of conservation of something in Nature, which was brought to that concept-creating activity which finally resulted in the development of the concept of energy: the work of Hermann von Helmholtz. In other words, concepts are in a state of flux while the discoverers of physical laws are working with them, or rather in them, and they become fixed only as a result of the law which after having been formulated mathematically, actually spells them out.

The personal acknowledgements will follow this preface, but here I wish only to note intellectual debts to Émile Meyerson, Pierre Duhem, Alexandre Koyré, Sir Karl Popper, Imre Lakatos and Robert K. Merton, whose works made a lasting impression on me, and to Mrs Haya Hillel to whom I am indebted for so much that I cannot express my debt in words.

Foreword

by *Professor I. Bernard Cohen, Harvard University*

Historians of science, philosophers and sociologists of science, and even scientists themselves today take it as axiomatic that historical examinations of science cannot be limited to the production of chronologies. There is general agreement that scientific thought should not be studied in isolation from the lives and characters of the men and women who have been its creators, and that philosophical examinations of science may profit greatly from the introduction of a historical dimension. An additional requirement put upon historical analyses of science is an awareness of the social contexts of scientific thought and scientific action and of the possibilities of sociological insight.

Having been nourished by such influential philosopher-historians as Léon Brunschvicg, Ernst Cassirer, Pierre Duhem, Ernst Mach, and above all by Alexander Koyré and Karl Popper—to list only a few of the most outstanding founding fathers of our generation we particularly demand a sound philosophical base for historical understanding. The theme of energy and its conservation, as discussed by Yehuda Elkana in this monograph, provides an admirable instance of the necessity for philosophical alertness in untangling the skeins of the history of scientific ideas. He demonstrates that many of the primary questions concerning the 'final' or 'definitive' enunciation of the theorem of conservation of energy cannot even be exactly formulated by the inquiring historian, much less answered, without preliminary philosophical analysis. The oft-debated issue of priority of discovery, or the alleged simultaneous and independent discovery, of the conservation of energy is clarified by Elkana on the basis of an acute philosophical analysis of the concept of energy itself, as it gradually emerged in the course of the nineteenth century. He demonstrates convincingly that this topic can be fully comprehended only by an awareness of the philosophical significance of the doctrine of conservation in general, both in the development of the physical sciences and in the scientific thought of the nineteenth century.

Despite the obvious Greek roots of the word 'energy', the emergence of a significant concept associated with this word dates only from the first decades of the nineteenth century; the modern sense of 'energy' in physics was defined in many stages, and required a dissociation from, and clarification of, other concepts such as 'force'. Although Thomas

Young consciously introduced the technical term 'energy' in 1807, this first 'modern' usage—as Elkana shows us—is limited to what nowadays we would call 'kinetic energy'; furthermore, the way in which Young made use of this term clearly rules out any possibility that he might have believed in, or conceived of, conservation as one of the properties of energy. It thus appears, from Elkana's analysis, that only after Helmholtz had established conservation as a primary property of energy, can we say that the concept of energy itself emerged in that bright clarification that was characteristic of physical science before the advent of quantum theory and relativity. It is thus, on the basis of a painstaking analysis of the meaning and physical significance of the basic concept, that Elkana demonstrates for us why the claims to discovery of the theorem (or 'law') of conservation of energy date only from *after* Helmholtz's foundational paper of 1847.

With respect to 'conservation', it may be noted that, of the four conservation 'laws' that came to dominate classical physics, the energy relation was the last to emerge, and was also the most short-lived. First, in order of time, was the law of conservation of momentum, a product of the seventeenth century (that is, linear momentum; the associated law for the conservation of angular momentum came considerably later). The mid-eighteenth century produced the law of conservation of charge, while the law of conservation of matter dates from the late eighteenth century or early nineteenth. The law of conservation of energy differs from the others in a very fundamental conceptual way. The conservation of momentum required a philosophico-scientific decision as to whether in collisions and free fall, the quantity 'mv' or '$mv.$' may be conserved, and also what feature or features of colliding bodies determine conservation (i.e., whether the collisions are elastic or non-elastic). A general guiding principle, however vague it may seem to us, that led to the concept of conservation of momentum was that there is a certain definite or finite quantity of motion in the world which can be transferred but never destroyed. This partly theological doctrine led a Descartes to an enunciation of the law of inertia and to the radical viewpoint that motion may be a 'state' rather than a 'process'; and it even led to a belief that motion, being conserved, may be converted into (and thus 'generate') heat. To paraphrase an argument of Robert Boyle, a moving hammer drives a nail into a board, converting the motion of the hammer into the motion of the nail; but when the nail-head is flush with the board, so that the nail can no longer move on into the board, the hammer-blows produce heat in the nail, for now the motion of the hammer is converted by successive blows into increased motion of the intestine parts of the nail, or heat.

The law of conservation of charge is of a somewhat different nature, for it is a conceptual generalization of the experimental observation that in the production of charges (as by friction), and in the neutralization of charges, negative and positive charges are either produced or disappear in equal magnitudes. Hence, a conclusion is that charge production is the result of a redistribution of one or two electrical fluids, and that the neutralization of charges results from a similar process. The conservation of matter may seem to be even more simply a generalization of experience, notably the observation that inside a sealed retort (or other vessel), there is no change in weight as the result of chemical action or reaction. That the formulation of these conservation laws differs basically from the circumstances of the conservation of energy will appear from Elkana's analysis. For in this case, a precondition for the concept of conservation of energy was to make precise the various forms and manifestations of energy, to analyse their interconvertibility, and to establish quantitative measures of energy. In distinction to the laws of conservation of momentum, charge, and matter, the law of conservation of energy required a high degree of philosophical *sagesse*, and accordingly a satisfactory understanding of its history must be based on a sound philosophical analysis of the developing concepts of energy which the law was to eventually fuse.

That all science should be based on conservation laws turns out to be a nineteenth-century ideal. Since today we can no longer believe in conservation of matter and of energy as independent and universal truths of science, we are at a loss to know whether to refer to the conservation of energy (and of matter) as a 'law' to which there are notable exceptions, or as a 'principle' or 'doctrine' of limited application. Physicists of the nineteenth century came to regard conservation as a major guiding principle of physical science, and made a near-religion of 'energetics'. But in our time physicists tend, as often as not, to be excited by situations in which conservation fails to hold (as in the case of 'parity') as much as by new instances of its success.

From ancient times until now, a guiding principle in physical science has been the search for constancies or regularities in a world of change, in an attempt to find order in a world of apparent chaos. If one does not accept the point of view that nature produces its effects by stochastic processes in a world run wholly by chance, then the search for regularities and laws must eventually lead to principles of conservation among the phenomena of change. As Elkana shows us, however, the 'scientific metaphysics', which was a feature of 'the intellectual environment of Europe in general, and . . . Germany especially', tended to 'centre around conservation ideas'. Here we have an example of the way in which the 'scientific metaphysics' of Helmholtz, and even his programme of research in the sciences was influenced by the general intellectual climate. Elkana goes a step further, exploring how the very choice of problems investigated by scientists may be related to, or influenced by, 'competing images of science'. In Helmholtz's case, this included a commitment to a general belief in the comprehensibility of nature and a concomitant conviction that natural events must be considered under organizing principles characterized by a general validity.

The study of energy and its conservation leads at once to certain sociological considerations. The development of the steam engine as a major source of energy not only presented important problems to scientists in a socially significant way, but also raised questions of genuine scientific importance. Elkana draws our attention, however, to another level of sociological inquiry—namely, the 'comparative historical sociology of scientific knowledge'. Although he modestly concludes his book by raising a series of questions for us in this area, questions which remain to be answered before the presentation of the doctrine of conservation of energy can be considered complete, Elkana has made a major contribution to the current dilemma of historians and philosophers of science, to the so-called internalist–externalist controversy. Displayed in the opening chapter, what Elkana proposes is a schema based upon a special set of interactions which occur between the scientific, the philosophical, and the social factors. This novel mode of conceiving the growth of science which analyses the hard core of experiment and observation and the concomitant concept formation in relation to the philosophical and the social environments, may well point the way to a wholly new level of understanding of the nature and nurture of science.

ACKNOWLEDGEMENTS

This book originated in a Ph.D. dissertation written under the supervision of Professor Stephen Toulmin at Brandeis University. I am grateful to him for having advised me to do history of science thirty years earlier than I intended to. When still in thesis form, Professor T. S. Kuhn read the work and was most generous with his time, and helpful with his broad knowledge of the subject. I thank him deeply. Needless to say that whenever I did not follow his advice, I knew the risk I was taking.

To my friends and colleagues: Y. Bar-Hillel, I. Bernard Cohen, Robert S. Cohen, Shmuel N. Eisenstadt, Erwin Hiebert, Gerald Holton, Imre Lakatos, Robert K. Merton and Arnold Thackray I owe many hours of fruitful discussion and guidance. Due to these discussions I now fully realize the shortcomings of this book. I wish to thank Professor I. Bernard Cohen especially for his interest in this work. To Mrs Haya Hillel I am indebted for so much that I cannot express in words. To my ever patient friends-secretaries, Mrs G. Shalit and Mrs J. Friedgut—gratitude.

Y. E.

I
INTRODUCTION

PHILOSOPHICAL BACKGROUND

General

Sir Karl Popper, in the Chairman's address to the Philosophy of Science Group of the British Society for the History of Science in 1952,[1] quotes Wittgenstein's 'Whereof one cannot speak, thereof one must be silent'; but he also quotes Schrödinger's famous reply: 'But it is only here that speaking becomes worth while.' Those famous last words of the *Tractatus* and the Schrödinger reply constitute the first thesis of Popper: 'Genuine philosophical problems are always rooted in urgent problems outside philosophy, and they die if these roots decay.' Problems that are considered urgent in science (and in other fields outside philosophy) are so considered on philosophical grounds, and are influenced by philosophy. In other words, there are no eternal philosophical problems, or eternal scientific ones. In every age some questions are considered purely scientific while others are delegated to philosophy. It is the interaction between the two which is perennial, and the most fascinating study. It is this interaction which helps us to see our own age in science and other fields in historical perspective; also it is this interaction which helps us keep in mind that every age has its 'modern' and its 'ancient' thinkers, its great riddles, its urgent scientific problems and those problems which it considers solved. Even the knowledge that scientific solutions are only temporary is not a twentieth-century invention: we find in every age great thinkers who profess to this temporariness; the most we can say is that some ages are more time-conscious and historically-minded than others. One is tempted to say that we live in an age which has completely absorbed the theory of evolution, and is indeed very history-conscious; but could this be said of the ruling school in philosophy and the ruling physical theory? It is disputable whether logical positivism is still the ruling school of philosophy (in the view of many it is on the decline), but there is no question about its anti-historical character. As to quantum mechanics, let me emphasize only that more and more important papers on new approaches to the impasse in this field deal with concepts of time.[2]

It was a glorious mistake on the part of Descartes to think that one could

solve the fundamental questions—those about the metaphysical foundations of natural science—and then let science build up its structure from
that point onwards. Modern science has learned not to pose those questions knowing that, if it did, it could not progress at all. Today's science
student is brought up on highly refined and sophisticated mathematical
and experimental techniques and has an enormous abundance of facts
to master, and in his curriculum there is no place for fundamental
questions. If he asks at all about the connection between those sophisticated techniques and the so-called 'physical reality' he is given a brief
exposé of the Copenhagen interpretation of quantum mechanics. If we
consider modern science and especially physics, in 1973, as a successful
enterprise, then this approach is justified and serves best; in this case
scientists are justified even in saying that in their everyday activity philosophy would only disturb them, not aid them. But is it an unquestion-
ably successful enterprise?

It has been different with regard to the few great steps in the development of our world picture: classical mechanics, relativity and quantum
theory (to mention only the most comprehensive classes). These developments followed the posing of the fundamental questions, and very
probably would never have taken place without having posed them. From
this point of view, history of science is not to be disentangled from philosophy of science, and the interaction between science and philosophy is
strong and very illuminating.

The interaction between science and philosophy takes place on two
different planes. On the one we find philosophy imbedded, or rather
rooted, in science, as in the cases of Kant, Schlegel, Fichte, even Hegel,
Comte, Mach and Poincaré. On the other plane it is the reverse: here
science is imbedded, or rather rooted, in philosophy, as in the cases of
Liebig, Johannes Müller, von Helmholtz, Faraday. Naturally, such a
dichotomy is oversimplified, yet it is basically true—when we compare
Descartes's system with that of Newton, nobody questions that basically
Descartes was a philosopher while Newton was fundamentally a scientist. It
is exactly this over-simplified dichotomy which I have in mind when
talking of two different planes. Kant was a philosopher whose problem
situation came from Newtonian physics; that was also the case, in my
opinion, with the 'Naturphilosophen'; Comte fed on Fourier's physics;
Faraday's metaphysical preoccupation is well known today; Mach and
Poincaré were in the main philosophers even if their activities were
often in the field of science. That this emphasis is not widely accepted at
least for the last two is, I believe, only due to the fact that these are still
near in time: in two hundred years they will be considered as philosophers
with minor contributions to science; Descartes is remembered for his
philosophy, not for his law of inertia or law of sines. (As for pure mathe-

matics, both Descartes and Poincaré fill a special place, but even here one could easily show the dependence of their mathematics on their philosophical principles.) On the other plane we have those great physicists whose major conceptual contributions were derived from having posed the basic philosophical questions: 'biology', which is a major conceptual development, was created by Liebig, Johannes Müller and their co-workers; the 'field' concept was born in Faraday's mind, and the concept of energy originated with Helmholtz. There were great contributions by many others: the French mathematical physicists, Maxwell and Kelvin, their physics was rooted to a much lesser extent in philosophy, and they provided new mathematical formulations or great syntheses, but rarely created new concepts.

Finally, such an approach as is given here carries with it, needless to say, a distrust of such historical studies that are written from the point of view of sudden revolutions. The slow mutual influence and interaction between science and philosophy presuppose a different kind of development. I shall skip here a review of the highly interesting philosophical discussion that has been going on in this field in recent years. I. Bernard Cohen, in his Introduction to the Wiles Lectures, says the following:

> Revolutionary advances in science may consist less of sudden and dramatic revelations than a series of transformations, of which the revolutionary significance is not realized (except by historians) until the last great final achievement. Thus the full significance of a most radical step may not even be manifest to its author.[3]

The traditional three great critical dialogues

The history of science since Newton, and due to Newton's success story, is often seen in the light of the debate between Newtonian and anti-Newtonian science.[4] But what is to count as Newtonian and what as anti-Newtonian? This is not a rhetorical question but an historiographical problem of fundamental importance, any reasonable answer to which would presuppose a comprehensive knowledge of the growth of science from the seventeenth century onwards. To describe the world in terms of discrete particles between which central forces are acting at a distance is certainly Newtonian—this is the metaphysical core of the *Principia*. The programme of mathematization of mechanics, as perfected by the French school of rational mechanics, is also Newtonian. The various matter theories of the eighteenth century first presented in the *Queries* and in the letters to Bentley, are correctly described by both Schofield and Thackray as characteristically Newtonian. The idea of chemical affinities is Newtonian but so was the Daltonian revolution which rejected the affinities. Lavoisier was a Newtonian of sorts and so was Priestley (whom Lavoisier rejected) so too was Humphry Davy, who refuted Lavoisier's central doctrine, namely that all 'elements' contain oxygen. Some of these great

natural philosophers called themselves Newtonians because they adhered
to a world view where the most important force was gravitation acting-
at-a-distance; some others accepted a material substratum, the ether, which
transmits all physical action and this was the reason why they considered
themselves Newtonians; some others again made the same claim because
they believed they were doing scientific work in the hypothetico-
deductive way, which they considered to be the hallmark of Newtonianism. Needless to say there is very little similarity between Newton's
thoughts and speculations and the conceptual framework which they
thought to be Newtonian. In addition to those who thought themselves
bona fide Newtonians, there are others who used the label 'Newtonian'
politically, for legitimization of their theories; an example is Thomas
Young, a disciple of Euler and Huygens, who introduced his famous paper
on interference of light by attributing the main ideas to Newton. Finally
there were the continental natural philosophers, all of whom accepted
Newtonian mechanics, and attempted to blend it into their Cartesian or
Leibnizian conceptual frameworks; later historians in their positivistic
whitewashing exercises called them Newtonian; such were, to name only
a few, Boscovich, Euler and Kant.

But, allowing for the moment when we could sufficiently refine our
notion of Newtonianism, it still seems to me that to attempt to describe
the main lines of the development of science in terms of the struggle of
the Newtonian and anti-Newtonian traditions is to put ourselves into a
conceptual strait-jacket.

In my opinion there were at least three great traditions or scientific
research programmes competing for primacy in science. These are the
Cartesian, the Newtonian and the Leibnizian research programmes. The
critical dialogue between these three was conducted in pairs: Newtonianism *v.* Leibnizianism; Newtonianism *v.* Cartesianism and again separately
Leibnizianism *v.* Cartesianism, or rarely when two joined forces against
the third. To lump all general explanatory hypotheses which are not
Newtonian together under the heading 'anti-Newtonianism' is an over-
simplification. Those conceptual frameworks which can justly be labelled
as anti-Newtonian focus their opposition either on Newtonian science or
Newtonian methodology. Yet anti-Newtonians proper and Newtonians
share a fundamental problem-situation: should or could one describe the
universe in terms of discrete particles with central forces acting between
them; can force act through a vacuum; are forces essential properties of
matter? On the other hand the eighteenth-century Leibnizians and the two
different brands of Cartesians (which separated out of the original Cartesian framework at the turn of the century)[5] had to face different problem-situations and had a different scientific research programme than the
eighteenth-century Newtonians. The two Cartesian groups were the

Cartesian mathematical rationalists like d'Alembert, Diderot and later Lagrange and the Cartesian matter-theorists like Maupertuis, Euler and Johann Bernoulli. Cartesian mathematical rationalism developed a programme aimed at subsuming all phenomena under mathematically formulated laws. Here there was no discussion of fundamental concepts, no search for underlying principles, and the criterion of truth was rarely empirical. Rather, mathematical formalizability and elegance became signs of truth. These Cartesians were occupied in developing mechanics as a branch of mathematics and concentrated on attacking the Leibnizians rather than the Newtonians.

The main argument between the matter-theorist Cartesians and the Newtonians centered on the primacy of the concept of force. These Cartesians too accepted Newton's results, that is, the laws of mechanics and the law of gravitation,[6] but they insisted that there are essential qualities of bodies to which forces can be reduced. If forces were introduced into the Cartesian programme they were considered as mathematical abstractions useful for smooth calculations—an attitude somewhat similar to that of Heinrich Hertz a hundred and fifty years later.

The mind-body dichotomy was part of the Cartesian tradition but it played only a very minor rôle in the controversy with the Newtonians. This problem was, however, the core of the Cartesian–Leibnizian critical dialogue. The Cartesians separated mind and body, and also scientific metaphysics (that is those views on the structure and genesis of the physical world which are in principle untestable, but form the core of their research programme) from theology. Both the Newtonians and the Leibnizians, on the other hand, attempted to justify their scientific metaphysics by their theology. This justification became one of the foci of the Newtonian–Leibnizian critical dialogue, as exemplified in the Leibniz–Clarke correspondence and as continued by Euler in the *Letters to a German Princess* written in the 1770s.

The central Newtonian conception is that of force, whether acting-at-a-distance or at short-range by contact. Newtonian physics, astronomy, chemistry and physiology all involve forces. Whether the forces are inherent in matter or reducible to their relational properties is another focus of the dialogue between Newtonians and Leibnizians. On the other hand, the concept of force is as foreign to the Cartesian as it is inseparable from both the Newtonian and the Leibnizian research programmes.

Another difference between Newtonians and Leibnizians is that conservation principles are alien to the former but fundamental to the latter. Even though an anti-conservation-principles attitude is not explicit in Newton's writing it seems to me to be one of his deep-seated anti-Cartesian biases. He, unlike Descartes, will not address himself to questions like—'what are fundamental entities?', 'are they conserved?', etc. He takes four 'fundamental notions'—space, time, mass and force—for granted and operates with them. For Leibniz too, the concept of force is fundamental, but it is rather its conservation which is at the core of his scientific metaphysics. It is the idea of conservation of force which served Leibniz in doing away with the Cartesian mind–body dualism and helped him develop his monistic theory.

In short, in order to gain any reliable picture of the growth of science one has to explore at least three competing traditions. All three left their indelible mark on the developments of science in the nineteenth and even the twentieth century; each at times had the upper hand in the long critical dialogues between them. Newtonianism is the paradigm of success in terms of positive scientific results. The positivistic attitude does not find a place for either the Cartesians or the Leibnizians in the history of science. Thus 'Newtonian' *v.* 'anti-Newtonian' covers the ground adequately only if we judge the development of science presupposing that science grows by accumulation. If we view the growth of knowledge as a result of a dialogue between competing research programmes we must think in terms of at least the above-mentioned three traditions.

How does knowledge grow?

An additional conceptual framework in which this book is imbedded is the following.

The central problem for historians and philosophers of science in the 1970s is: 'how does knowledge grow?', i.e. what causes change in the contents of knowledge and 'what is that part which "grows"?', i.e. serves as the nucleus of accumulation and continuity. My approach to this problem is based on the following theses.

All analysis of changes that cause the growth of knowledge has to rely on three different kinds of interacting factors:

(i) Developments in the body of knowledge which emerge from the scientific ideas themselves and which point to possible directions of change. Such developments grow out of the scientific metaphysics which is at the core of the body of knowledge, or, as Lakatos pointed out, of a Scientific Research Programme, or (in Merton's language) of a 'Strategic Research Site'.[7] Scientific metaphysics comprises statements which are about the structure of the world but are formulated in such terms that they are not directly testable—neither confirmable nor refutable. Newton's scientific metaphysics was that the world consists of discrete particles between which central forces (and such only) act by way of attraction or repulsion. Faraday's scientific metaphysics was that the world consists of a plenum of forces and that forces are convertible into each other while their total quantity is conserved. The scientific metaphysics of modern molecular biologists is a severe reductionism (i.e. the belief that *all* biological phenomena are reducible to *known* laws of physics and chemistry). These metaphysical statements are scientific inasmuch as they concern the structure of the world, and one can argue about them rationally, unlike ethical or religious metaphysics. Indeed, the history of the growth of knowledge is the history of critical dialogues between competing scientific metaphysics, that is competing scientific research programmes.

The most important characteristic of disembodied pure knowledge is that here all developments satisfy the necessary conditions for a new discovery. However when the discovery is actually made by which open problems are taken up, what is considered the 'frontier of knowledge' is decided by the sufficient conditions of the 'image of science'.

(ii) The social image of science (or of knowledge in general) in a given place or community at a given time: whatever people in general and scientists in particular *think* consciously of science, its rôle, its ethos, etc. This influences heavily the choice of problems from among the enormous range of open problems as provided by the body of knowledge itself (i.e. by factor (i)). The image of science shapes the formulation of selected problems, determines what is called the 'frontiers of science' and determines the *reasons* for scientists' promotions. It is sometimes so influential that it not only emphasizes some of the open problems but even totally obscures others. It decides what is legitimate science and what is pseudo-science; it shapes the demarcation criteria between science and metaphysics. The image of science has its own metaphysical part in what was usually called 'Zeitgeist' or 'Weltanschauung' and its sociological-pragmatic element. What is common to both is that they are on a cognitive level.

(iii) Social and political factors which interfere directly with the lives of the scientist and the scientific institutions and thus influence the development of science via direct manipulation. To this category belong such trivial examples as the fact that in Soviet Russia there were no Mendelian geneticists for twenty years because they had been liquidated one way or another. A more sophisticated example would be that the active royal support of the Académie Royale des Sciences made its members much more interested in the practical needs of the state while the early members of the Royal Society disregarded such questions.

These three kinds of factor interact. Social and political factors in addition to influencing directly the lives of the scientists, also influence the image of science, i.e. what people *think* of science. All scientific metaphysics is heavily influenced both by developments in science and by the cultural and social environment with which it is intimately linked.

On the other hand new scientific ideas, insights and products certainly influence both the image of science and the socio-political developments. These interactions are so interpenetrating that one can easily draw the irrational conclusion that no analysis of factors can be undertaken and rather look for hidden personal motives and for the one, underlying, unifying principle which makes humanity tick. This conclusion is so strange to all rational scholars that they prefer to rush into absolute dichotomies like science *v.* metaphysics, or internal *v.* external history, and write 'a rational reconstruction' which they admit has very little to do with actual history.

My justification for the above analysis into three kinds of interacting factors is that it seems to me to be much more satisfactory than the internal *v.* external historical explanation and that it obviates the necessity of a demarcation criterion between science and metaphysics; it helps to realize a reconstruction of the changes in the past which is both rational and historical.

Much as these factors interact with each other historically, at any point in time a horizontal cut can be made; one can then easily distinguish between the 'internal' body of knowledge, the social image of knowledge and the social rôle of the man of knowledge and between the socio-economico-political 'external' factors. For example in modern physical science the most advanced research consists of investigation into elementary particles, the solid state, and field theory. The image of science is that science has very little to do with physical reality, and its task is to predict the world—as long as the theory provides correct predictions the theory is good; otherwise it is replaced. The socio-political constraints are the distribution of grants which heavily encourage the research which promise technological and military possibilities while far less research is encouraged to deal with fundamentals like a possible revision of quantum mechanics, or unified field theory.

How can one support this thesis? I do not know of any way other than that of the historical inductivist: to pile up case histories and to allow the burden of evidence to carry its own weight.[8]

THE PLAN OF THE ARGUMENT

The central concepts in Newtonian physics were space, time, mass and force.[9] By the end of the nineteenth century the central concepts had become space, time, mass and energy. My aim is to trace the emergence of the concept of energy; to investigate the historical, philosophical and scientific factors that brought the above-mentioned change; and to analyse the relationship between the concept of energy and that of force, in view of the confusion which ruled in the nineteenth century.

The general concept of energy became meaningful only through the establishment of the principle of conservation of energy in all its generality. Thus the story of the emergence of the energy concept and the story of the establishment of the conservation law are difficult to disentangle; they are intimately connected while there exist numerous[10] histories of the law of conservation of energy.

It is admitted by all, that the man who formulated the principle for the first time mathematically, in all its generality, was Hermann von Helmholtz, and thus necessarily this essay on the emergence of the energy concept will centre around him. His was a towering scientific personality and his lifework has left its marks on all branches of nineteenth-century science; from theoretical mechanics to applied physiology. It will be claimed that the concept of energy as we know it today (by 'today' classical, pre-relativity physics is meant) has emerged from Helmholtz's 1847 paper 'Über die Erhaltung der Kraft'[11] and that up till then, nobody, including Helmholtz himself, had a clearly defined concept of energy.

But the problem is not a purely scientific one. Helmholtz's part in this story is unquestioned—it has never been questioned. But the claim is further than that: in view of the prerequisites of the principle of conservation of energy, which will be outlined further on, it is natural that this final step took place in nineteenth-century Germany, and not in England or France. (As far as the nineteenth century is concerned, it is enough to consider only England, France and Germany: naturally, individual geniuses could have sprung up, and in fact did, in many other countries.[12])

It will be shown that the confusion between 'force' and 'energy' (as we use these terms) in the works of Helmholtz and some of his contemporaries, was not only a verbal one, as most of the commentators on this topic tend to assume, but rather a necessary prerequisite for the final clarification of the concepts. Only an undefined entity could have been the subject of a general belief in principles of conservation in nature, and as I will try to demonstrate, such a belief was one of the major factors in the actual establishment of the conservation of energy principle in its final and mathematical, that is, correct and well-defined, form.

I will examine some of the often quoted 'roots' of the principle of the conservation of energy, and will attempt to prove that while some of them were either secondary in importance, or had no rôle to play at all, some others which had been generally neglected, were the most important factors leading towards the final formulation of the principle and the emergence of the 'energy' concept.

One of the most cherished beliefs of many historians of science is that the principle of conservation of energy grew directly out of the realization of the impossibility of a perpetual motion machine.[13] This realization is indeed one that was arrived at inductively, and it dates back to the seventeenth century if not to earlier times. Certainly Stevin had already drawn physical conclusions from it. In 1775 the French Académie des Sciences declared that no more suggestions for the construction of such a machine would be considered, not only in mechanics but in *all branches of physics*. Thus at least seventy years before the establishment of the conservation principle, the impossibility of a *perpetuum mobile*,[14] the alleged 'intellectual father' of the law of conservation, had been established beyond doubt. It will be shown that although Helmholtz himself claimed this ancestry for his proof, it was far from being a sufficient condition for it. Otherwise in 1775 the conservation laws could have been formulated without any hindrance. At best it is a necessary condition insofar as it is implied by the principle of conservation: if the principle of conservation holds, indeed no perpetual motion machine is possible. The principle of conservation of energy is not the result of a long inductive process, but a 'science-producing' presupposition (an expression of Maxwell's).[15] To what extent this is true will be underlined by the story that William Thomson (later Lord Kelvin), while working on Carnot's theory and approaching the subject with the impossibility of a *perpetuum mobile* in mind, at one and the same time lectured on electromagnetic induction, assuming without hesitation that in this case mechanical energy is simply lost without any compensation.[16]

It is also often assumed that the principle of conservation of energy was a direct generalization of the law of conservation of mechanical energy, as formulated, for example in Lagrange's *Mécanique Analytique*.[17] But here, the use of the modern name of mechanical or kinetic and potential energy leads us into hindsight; it is implicitly assumed that the concept of energy was extant, and that the workers in this field thought in these terms and worked with this notion. Actually it is only now that we view all sorts of work, '*vis viva*', 'Kraft', 'Spannkraft' and the many others, as instances of the all-embracing concept of energy. At the time, when in mechanics the sum of '*vis viva*' and of 'potential function' (under this or any other name) was found to be conserved, nobody thought of the necessity or possibility of generalizing this any more; it was general enough. The notion of energy, as something so general that all the special forms are only instances of it, was created only in the late 1840s, and then in precise mathematical language. The importance of the mathematical language must be emphasized because long before 1847, the concept of energy was implied as a vaguely understood identity which was conserved and related to mechanical energy, and even served as a working concept very fruitfully; it was used by Faraday, Mayer, or any of the twenty-odd 'simultaneous discoverers'[18] of the principle. Among the considerations of Helmholtz was that due to exact mathematical and dimensional considerations the conserved entity had to be related to mechanical energy by simply being reducible to it.

One often finds in the historical or even physical literature that at the beginning of the nineteenth century there were still two theories of the nature of heat in vogue: one was the mechanical, or rather, dynamical theory (as the name 'mechanical' is really justified only after the work of Clausius) and the other the material-caloric theory. It is implied or sometimes explicitly stated that Carnot had a clear conservation law in mind and was only misled by the use of the caloric theory, and that the mechanical theory had to be established in order that the principle of conservation could finally be enunciated. As to this, it will be shown that the connection between the actual development of the early thermodynamics and the theory of the nature of heat was very weak; that even a year after Helmholtz's proof, some scientists held to the caloric theory of heat (like Clausius, although he must have known of Helmholtz's work); that the conversion processes which were available due to the work of many physicists from Carnot to Joule, did not really point to a general conservation law. This again would seem to us natural because we already view heat as one of those many instances of a general concept of energy. The early famous supporters of the mechanical theory—Rumford and Davy— did not entertain any conservation ideas. Rumford actually showed the exact opposite: his argument was that, the heat generated being inexhaustible (that is, clearly not obeying any conservation law) cannot be material (as material substances do obey a conservation law); Davy, if his experiment teaches anything at all (which was cast in doubt by Andrade,[19]) has nothing to do with conservation. In short, the historical development was:[20] first the establishment of the principle of conservation of some entity (perhaps force, perhaps something related), and through it, the emergence of the concept of energy; after that the formulation and separation of the two laws of thermodynamics; finally the mathematical formulation of a true, mechanical theory of heat. It was only after these developments of thermodynamics that the actual processes of nature took an important place alongside the possible ones and a new interest arose in the extremum principles which now became an integral part of the new energy-centred mechanics.

The following are the factors which constituted a solid basis for the enunciation of the conservation principle.

(1) An *a priori* belief in general conservation principles in Nature.

(2) Realization that it is not enough that the two formulations of mechanics—the vectorial–Newtonian and scalar-analytical–Lagrangian—are mathematically equivalent; they must also be correlated conceptually.

(3) An awareness of the physiological problem of 'animal heat' or more generally of 'vital forces', and a belief that these are reducible to the laws of inanimate nature.

(4) A mathematician's certainty that whatever is the entity which is conserved in Nature it must be expressible in mathematical terms, and a mathematician's skill to perform the task.

That the *a priori* belief in general conservation principles in nature and the cross-fertilization between the physical and physiological sciences was a typical characteristic of the German university education of the time, will be dealt with in a separate chapter. Hermann von Helmholtz happened to be one scientist who combined in himself all those necessary requirements.

Helmholtz had read at a very early age the works of Newton, Euler, d'Alembert and Lagrange (though not Hamilton); was aware of the double tradition in mechanics, that the central concept in Newtonian-vectorial mechanics was the concept of force, and that no conservation principles formed a basis of the Newtonian conceptual framework. At the same time the quantity conserved in scalar–Lagrangian mechanics was the sum of '*vis viva*' and the 'potential function'. By temper and intellectual heritage he was a disciple of Kant and thus committed to a belief in the great unifying laws of nature;[21] this took the form of conservation laws, and naturally the conserved entity had to be that vaguely defined entity 'Kraft' ('force' in the Faraday sense). All this was in complete harmony with his mechanical philosophy: a belief that all phenomena of Nature are reducible to the laws of mechanics. By training he was a physician and he spent several years in the laboratory of the famous physiologist Johannes Müller. There he came to face the problem of 'vital forces' and especially that of animal heat, and his first works were in this field. Again, his argument was that 'vital forces' are like other forces, conserved in Nature and as all phenomena are reducible to mechanics, so 'vital forces' must be reducible to mechanical forces. On top of all that, Helmholtz was a mathematician of the first rank. He saw very clearly that if 'Kraft' is conserved in Nature, and mechanical energy is conserved in mechanics, then all 'Kraft' must have the same physical dimension as mechanical energy and must be, moreover, reducible to it. That is exactly what he did in his 1847 paper. These lines of the argument will be taken up one by one in the following chapters.

THE GENERAL POINT OF VIEW

What is described above is the historical aim of this study, and this being a historical study it is its more important aim. On the other hand for philosophical frankness it must be stated that exactly as I uncompromisingly reject the 'blank-mind-experimenter' theory, namely, the view that it is possible to do science without a preconceived theory which one tries to test out, so also I reject the idea of the objective, 'blank-mind' type of historian who approaches the subject without any beliefs or any more or less clearly formulated historical conceptions which he wants to check. Naturally, one is justified in discussing and analysing the degree and extent to which the theory to be tested is clearly formulated, both in science and history; the theory I have in mind is not of the sort which would claim that the exact result of this or that experiment is clearly predicted in all its details, or that having been committed to the belief that this or that scientist is the *first* to have discovered the principle of the conservation of energy, now you go and write his biography. But I certainly do claim that no concept-forming theory of science ever grew out of a non-committal attitude to a specific model, even if the discovery made in the conceptual framework of one theory leads to the rejection and final abandonment of that very framework. Let me cite an example to make my point clear. It is well known that in his posthumous notes Carnot took up the motion theory of heat. His original memoir was an outgrowth of his conceptual commitments to the caloric model of heat. However, having arrived at a vague formulation of what we call today the second law, he realized the absolute necessity of a conservation law too, and in his posthumously published later notebooks he switched to the motion theory of heat.

I chose this example deliberately because it is not a case which is worked out here, and therefore shows more clearly the bias in which the studies which are examined here have been grounded. To make it even clearer, let me show my prejudice in a counter-example: I do not think that those great scientists who were engaged in mathematical syntheses, or gave differing mathematical models on the basis of contradictory physical models, ever created new concepts of physical science. This is not to say that these or the others are good or bad—this should not be a value judgement: the synthesizer-mathematicians are as indispensable as the concept creators; these and their approach constitute steps in the development of the sciences, which replace each other towards a more and more developed stage of the science in case. If I have to answer the question: 'If these are not value judgements, then what are they?', I shall have to answer: 'These are statements on different types of scientific thought' or 'different scientific tempers'—in short, statements which are fundamentally psychological in character. What are the 'concepts' which fit this dichotomy? I can give no cut-and-dried answer. None of those who talk of 'conceptual revolutions', 'conceptual evolutions' or 'conceptual frameworks' really try to define them. Yet, it is perfectly clear that what is meant by 'concepts' are those physical entities in terms of which scientists think when developing new mathematical formulae or plan new experiments. Concepts in which a scientist thinks must not necessarily be related to a mechanical, visualizable model; it means only that the concept must be accompanied by a content, which cannot be formulated in words. These words do not have to be defined clearly and unambiguously: it is the loss of the history of science and of the psychology of discovery that the code of behaviour, dictating the behaviour of scientists, prevents their expressing themselves vaguely and tracing the muddle in which their greatest thoughts have been formed. Classical exceptions to this rule are Leibniz and Faraday.[22] A notable modern exception to this is Richard Feynmann who, in his Nobel Lecture, and in other articles since, has tried with great honesty to explain and trace the conceptual origins of his discoveries. Mathematics is in itself neither conceptual nor unconceptual—the question is whether the mathematician can give any verbal content to his symbols. Here again an important qualification is necessary: I deal here only with the mathematics as it is applied to physics. On a completely different level, one can talk of conceptual or non-conceptual thinking in pure mathematics too; whether there too the criterion would be the possibility of formulating the content of the concept verbally, however vaguely, I do not know. Whatever the idea of a concept is, no one will doubt that 'energy' is a physical concept, which was created by conceptual science-creation, is intimately connected with the principle of its conservation, and fulfilled a very important role in consolidating classical physics, before it again received new content through the developments in relativity theory and quantum mechanics. I hope that I have made my approach sufficiently clear, and that I have committed myself completely. This general attitude will be applied to the case of Helmholtz.

THE POINT OF VIEW APPLIED TO THE CONCEPT OF ENERGY

I shall treat Helmholtz's fundamental paper 'On the Conservation of Force', as an illustration of what seems to me to be a general feature of how scientific concepts develop, namely, that they are in a state of flux while the individual scientist is struggling to clarify his thoughts, that is, while the discovery is being made. I suggest that Helmholtz believed in a vaguely formulated conservation law, and that only after his proof of the conservation of 'Kraft' had been accomplished did his concepts of energy and force (in our sense of the words) become separate and fixed. In order to see the problem from a historical perspective let us look at the contemporary opinions about Helmholtz's use of the word 'Kraft'.

Maxwell, in an article in *Nature* in 1877, says:

There can be no doubt that a very great impulse was communicated to this research by the publication in 1847 of Helmholtz's essay 'Über die Erhaltung der Kraft', which we must now (and correctly, as a matter of science) translate 'Conservation of Energy', though in the translation which appeared in Taylor's scientific memoirs, the word 'Kraft' was translated as 'Force', in accordance with the literary usage of that time.[23]

Maxwell could write that easily, as until this time the English scientific journals still used the word 'force' for 'energy' very often. But even as late as 1895 the 'substitution' approach was widely accepted. (By this I mean that it was thought that one could simply substitute 'energy' or 'force' for 'Kraft' as the case required on our terms.) Also, in the same year T. C. Mendelhall wrote in an essay on Helmholtz:

Its excellence is shown by the fact that if rewritten today it would be changed only a little in its nomenclature.[24]

Such examples could be quoted by the hundred. However, in note 59 to his 'Towards a Historiography of Science' Agassi writes:

Helmholtz, in his preface to the German edition (1870) of Tyndall's *Faraday as a Discoverer*, pooh-poohed Faraday's speculations, viewing them as a 'disadvantage' excusable in view of Faraday's 'want of mathematical culture'. He also blamed him for having misunderstood the law of conservation of force. Later, in his Faraday lecture (1881) Helmholtz unscrupulously changes his tune and also relabels his 'On the Conservation of Force' as 'On the Conservation of Energy'. In his Faraday lecture Helmholtz obliquely claims priority over Faraday of having advocated the law of conservation of energy. The first motive which guided him seems to have been an instinctive foreboding of the law of conservation of energy, which many attentive observers of nature had entertained before it was brought by Joule, to precise scientific definition. He, by contrast, like Mayer, Grove and Joule, followed Faraday in advocating a conservation of Force not of Energy.[25]

Agassi has been quoted at length, because, though I do not at all agree with his interpretation, he seems to be the only one who is clearly aware of the difficulties involved in the choice of words here, and sees that we are dealing with an undefined entity. Above all I do not agree with his remarks about Helmholtz's 'unscrupulous' *volte-face* in his terminology. Not only was Helmholtz famous for his scrupulous honesty, but it is exactly this change from 'Erhaltung der Kraft' to 'Constanz der Energie' which is not accidental, but is not conscious either; it must be shown here that this change was natural and inevitable.

Let me make it very clear that I am not engaged in any sort of historical debunking, nor do I want in any other way to detract from the value of

Helmholtz's contributions to science. To repeat, I consider this case as an illustration of that general aspect of scientific change which, to make it 'thought-provoking by the image it creates', I called concepts in flux. By this I mean, that although Helmholtz has proved correctly and generally the law of conservation of energy, at the time (1847), the two concepts (or rather what are for us the two concepts of force and energy) were not at all clearly defined separate entities, but rather different guises of a vaguely defined 'Kraft' which was being conserved. In other words, while great scientific discoveries are being made, the very concepts which the discoverers think in, or work with, are in a state of flux; only after they had completed their discovery and formulated the result mathematically, does the new concept emerge and become fixed; this fixing of the concept carries with it already the results of the last discovery out of which it emerged. For our case it meant that the concept of energy, as it is fixed in our minds now (the 'now' covers classical, pre-relativity physics) implies the conservation of energy principle as formulated by Helmholtz and others, and it is not the case that these discoverers had a clear idea of the energy with which they worked, and of which they tried to prove the conservation. To recreate the conceptual framework in which those people thought before the 'new' concept became fixed, requires not only historical precision but also an effort at psychological empathy. Even today we encounter conceptual difficulties working in a relativistic conceptual framework, because our language, and thus probably our thought-world too, reflects Euclidean geometry and Newtonian space-time structure in the same way as our thought-world incorporates the conservation of energy principle. As a result of the rapid development of 'energetics' in the late nineteenth century, this principle penetrated into philosophy and psychology on all the levels.[26] When we talk of 'kinetic energy', 'potential energy' or of heat (though to a lesser extent, as this requires still some degree of scientific background) we think of them as instances of one entity—energy. For the laymen and even for the physicists of the eighteenth century and nineteenth century this was not the case. It took till the time of Lagrange and his 'final' formulation of analytical mechanics for the mechanical philosophers to realize that even in mechanics the '*vis viva*' of a moving particle, and the various other concepts used, were one and the same. Even later, it was not by accident that what the Germans called 'Leistung', the English and some other Germans called 'Effect of force', Smeaton called 'mechanical power', Lazare Carnot called 'moment d'activité' or sometimes (for different cases) 'force vive virtuelle', etc.[27] All this simply means that when different processes were investigated—collision of elastic bodies, collisions of 'hard'[28] bodies, springs, inclined planes—mostly the conserved quantity was thought to be a different one. There were all the time philosophers of nature who talked about some vaguely conceived 'forces' which were conserved in nature; but it took Helmholtz to combine such an *a priori* belief with thorough knowledge of physics and mathematics and to realize what mathematical form this conserved entity must take.[29]

For the case of Helmholtz an attempt will be made to show that the concept in flux, that is the hammering out of the new concept while the discovery is being formulated, does not contradict the possibility of an exact mathematical formulation; that Helmholtz's background and the various intellectual influences on him support this approach, and that the internal evidence in the text of his paper shows that the very ambiguity in the word 'Kraft' is indispensable for the comprehensibility of the work. By intellectual influences I mean the double tradition in mechanics (the Newtonian–vectorial and the Lagrangian–analytical), the cross-fertilization between physics and physiology which in his case was very strong, and the philosophy of Kant. It is more difficult to analyse exactly what is meant under 'background'. This involves external influences and the image of science. In the nineteenth century there was a strong realization that the intuitive primitive idea is that of 'force'; that this concept is as ancient as human consciousness, and that as such it is necessarily vague. The usual climate of opinion is very clearly formulated by Moritz Schlick in his *Philosophy of Nature*:

The concept of force undoubtedly owes its origin to the muscular effort which human beings experience when they attempted to set bodies in motion.[30]

Here are a few expressions in which the eighteenth- and nineteenth-century scientists used the word 'force': 'force of a muscle', 'force of a machine', 'force of gravitation', 'electric forces', 'magnetic forces', 'galvanic forces', 'mesmeric forces', 'vital forces', 'forces of nature', not to mention all those expressions which had a theological overtone. These expressions have their parallels in all the other European languages. Naturally, all of those natural philosophers, who in the Cartesian–Leibnizian tradition[31] sought for general principles in Nature, felt very strongly that something in Nature had to be conserved. The Principle of Conservation of Force is that vaguely formulated principle to which Helmholtz was committed and which, as we shall see, together with the other influences, led him to a generalization of the conservation law of *vis viva* and which resulted in the creation of the concept of energy. From the 1880s Helmholtz himself talked of 'Constanz der Energie'.

In order to carry this point, one has to act on Einstein's advice:

If you want to find out anything from the theoretical physicists about the methods they use, I advise you to stick closely to one principle: don't listen to their words, fix your attention on their deed.[32]

There is ample evidence in Helmholtz's works to justify us in disregarding his proclaimed, extreme inductivism, and to work out the implications (though very briefly and incompletely) of the admitted influence of Kant on Helmholtz, namely Kant's mechanistic, causal theory and the connection of this with the conservation principle. This will be outlined in Chapter VII, on the philosophical parentage of Helmholtz's physical ideas.

NOTES

1. Reprinted in *Conjectures and Refutations* under the title 'The Nature of Philosophical Problems and their Roots in Science' (London, 1963), p. 66.

2. Special reference is made here to the paper by D. Bohm, 'Space, Time and the Quantum Theory' (1966).

3. I. Bernard Cohen, The Wiles Lectures, held at the Queen's University, Belfast, in 1966 (to be published).

4. For this argument applied in greater detail specifically to the eighteenth century see my 'Newtonianism in the eighteenth century', *Brit. J. Phil. Sci.*, **22** (1971), 237.

5. J. Rohault's *Traité de Physique*, which appeared in 1671 in Paris, is probably the last Cartesian work before the above-mentioned separation.

6. However, they did not accept Newtonian optics, but rather fought the Newtonians on the issue of essential qualities and the nature of light jointly with the Leibnizians.

7. On all this see Lakatos, Merton and Elkana.

8. I have tried three such case histories in addition to the present book.

The first concerns the discovery of conservation of energy in Germany by Mayer, von Helmholtz and others and the discovery of the motion theory of heat in England by Joule, Thomson and others. These two discoveries at first (because of differences in scientific problems, in the image of science, and in the socio-cultural setting) looked unconnected until in the 1860s they were conflated into one major discovery; see my 'The Conservation of Energy: a case of Simultaneous Discovery?' *Arch. Inter. d'Histoire des Sciences* **90** (1971), 31–60. It is reprinted here as an Appendix, p. 175.

The next case history is 'Boltzmann's Scientific Research Programme and the Alternatives to it', in Y. Elkana (ed.), *The Interaction between Science and Philosophy* (Humanities Press, 1974).

The final case history has the following historical thesis. Euler was a Cartesian in his metaphysics, a Newtonian in his methodology; his image of science was heavily influenced by Leibniz and by the tone of the Enlightenment as it was expressed by the Berlin and St Petersburg Academies, and he was a major influence over Immanuel Kant. On this see my 'Scientific and Metaphysical Problems: Euler and Kant' *Boston Studies Phil. of Sci.* XIV (1974), pp. 277–306.

9. It would be natural to include 'motion' among these central concepts. I left it out in order to emphasize the concept of force in that sense in which it replaces the concept of motion for both Newton and Leibniz. Descartes did not choose the concept of force but rather extension as a primary property. Much of the early criticism of Descartes was aimed at pointing out just this inadequacy in his system. Newton, though he defined 'motion' among the first concepts which he defined in the *Principia*, took the fundamental character of 'force' for granted.

10. Max Planck, *Das Prinzip der Erhaltung der Energie* (Leipzig, Berlin Druck u.

Verlag B.G. Teubner, 1913). This book was written in 1887 for a prize competition announced by the faculty of philosophy of the University of Göttingen. It constitutes one of the most extensive treatments of the history of the conservation principle and also a detailed exposition of its importance in science and philosophical meaning. It is interesting to note that Planck's treatment is very usual for its time in so far as it participates in the inductivist witch-hunt against the last remnants of the 'Natur-philosophie'; on the other hand his insights and explanation are far beyond any methodological school.

Other books are by Mach, Hiebert and Theobald, and the long paper by T. Kuhn. About these see below.

11. 'Über die Erhaltung der Kraft', read before the Physical Society of Berlin on 23 July 1847. As I shall be quoting from this work extensively, a few remarks about it will be in place. It appeared in the same year, published privately by Helmholtz in Berlin, Druck u. Verlag von G. Reimer. The essay is 53 pages in length and is sub-divided in the following way:

> Introduction
> (I) The principle of conservation of *vis viva*
> (II) The principle of conservation of force
> (III) The application of the principle to mechanical theorems
> (IV) The force-equivalent of heat
> (V) The force-equivalent of electric processes
> (VI) The force-equivalent of magnetism and electromagnetism

12. The overwhelming contribution of these countries, and the general assessment of it, is convincingly presented in J. T. Merz's *Introduction* to *A History of European Thought in the Nineteenth Century* (4 vols., Dover, 1965).

13. I refer here to the extreme empiricist-inductivist-positivist attitude which was the shared credo of nineteenth-century physicists. In this view the principle of conservation of energy is a conclusion drawn from the repeatedly unsuccessful attempts to construct a perpetual motion machine. As we shall see later, in reality they were only paying lip service to this philosophy, for science is not made this way. On the other hand, a great part of twentieth-century history of science is written from this point of view.

14. This will be treated in detail in the section on the *perpetuum mobile* issue (p. 28).

15. J. C. Maxwell, *Matter and Motion* (Dover edn), p. 54. There he says: 'As a scientific or science-producing doctrine, however, it is always acquiring additional credibility from the constantly increasing number of deductions which have been drawn from it, and which are found in all cases to be verified by experiment.'

16. Lord Kelvin, *Collected Papers* (Cambridge, 1911).

17. J. P. Lagrange, *Mécanique Analytique* (Paris, 1788).

18. I refer here to an article by Professor Thomas Kuhn, 'Energy Conservation as an Example of Simultaneous Discovery', in Clagett (ed.), *Critical Problems in the History of Science* (Wisconsin, 1955), and to my reply, see note 8 above.

19. E. N. da C. Andrade, 'Two historical notes', *Nature* (9 March 1935).

20. It is also the logical one. Rational reconstruction and historical narrative cannot differ in principle. On this see I. Lakatos and my reply, see note 8 above.

21. This is generally not considered the distinguishing element of Kantian thought: therefore see my *Kant* and *Kant & Euler*.

22. J. Agassi, *Faraday*.

23. J. C. Maxwell, 'Scientific Worthies: Hermann von Helmholtz', *Nature* xvi (1877), 389.

It seems to have been a characteristic of Maxwell to attribute to other great physicists only opinions which to him seemed the correct ones. We shall see below an even more convincing example of this attitude for the case of Faraday, where Maxwell assumes that Faraday knows exactly what the difference between 'force' and 'energy' is and only uses a confused terminology. But there, Faraday was conscious enough of the unusual and deeply-rooted, a-prioristic conservation principle in which he believed, to have corrected Maxwell, thus not leaving us in any doubt. The reason for this is, in all probability, that the very much 'mechanical-philosopher'-minded Maxwell, who did not have a-prioristic beliefs of this generality, could not at all understand that Faraday had vague notions which indeed were the source of his greatest contributions to science—the 'lines of force' and the 'field' concepts. And this much could be said not only of Maxwell but of very many great scientists—probably only those whom I would like to call 'Newtonians' in the spirit. This was certainly true of Planck, who, a few passages after having described Descartes's and Leibniz's 'vague ideas' about the nature of '*vis*', goes on to read Helmholtz correctly (see below). As an opposite example one could cite Einstein's tracing back his own thoughts to those of Mach and others. Naturally, it is no accident that Einstein was the most explicit 'anti-inductivist' scientist of our century.

24. T. C. Mendenhall, 'Helmholtz', *Ann. Rep. Smiths, Inst.* (1895), 787.

25. J. Agassi, 'Towards an Historiography of Science', *History and Theory*, Beiheft 2(1963).

26. The link between the two can be traced to G. T. Fechner, but nowadays we find it in the indiscriminate use of the various 'psychic energies', especially as used in Freudian theory.

27. In Chapter III on the mechanical traditions it will be shown that those mechanical philosophers who developed the mechanical energy conservation principle mostly saw one typical case which they treated paradigmatically and thus invented names for the conserved quantities in any one kind of experiment, not realizing that all of them were treating the same quantity. They certainly knew of each other's work and there was no necessity for such a diversity in terminology, but for the fact that they did not realize that they were all working on the same problem.

28. On the problem of 'hard' bodies see below.

29. I am fully aware that this theory of concept-formation, whether read primarily on a logical level or rather on a historical level, has to be demonstrated in many cases and more fully documented to carry conviction. I certainly hope to do so at a later stage. But in the meantime I have put it in a nutshell here in order to make clear the point of view from which Helmholtz's work will be analysed.

30. Moritz Schlick, *Philosophy of Nature* (Philosophical Library, New York, 1949), trans. A. von Zeppelin. The same idea occurs in Planck's *Geschichte*.

31. By Cartesian or Leibnizian in juxtaposition to 'Newtonian'—I refer to the three great critical dialogues mentioned in the Preface. Even if one does take into account all the qualifications and reservations that Gillispie mentions in his Introduction to Fontenelle's *Eulogium of Newton,* there is still basic truth to be found in those lines. Comparing Newton with Descartes, Fontenelle says:

These two great men, whose Systems are so opposite, resembled each other in several respects. But one of them taking a bold flight, thought at once to reach the Fountain of All Things, and by clear and fundamental ideas to make himself master of first principles; that he might have nothing more left to do, but to descend to the phenomena of Nature as to necessary consequences; the other more cautious, or rather more modest, began by taking hold of the known phenomena to climb unknown principles, resolved to admit them only in such manners as they could be produced by a chain of consequences.

The quotation is taken from *Isaac Newton's Papers and Letters on Natural Philosophy*, edited by I. Bernard Cohen (Cambridge, 1958), p. 427.

There is no need whatsoever to understand Fontenelle as if he described Newton as a primitive inductivist. Nobody takes seriously today the theory that one could enter a laboratory with a blank mind and 'do empirical research' for some time. Already in 1907 Meyerson says in the Introduction to his *Identity and Reality*:

> It must not be forgotten, indeed, that research is always dominated by preconceived ideas—that is, by hypotheses; contrary to what Bacon believed, those are indispensable in guiding our advance.

Or again Pauli, in his essay 'Phenomänen und Wirklichkeit' which appeared in the collection *Aufsätze und Vorträge über Physik und Erkenntnistheorie* (Vieweg und Braunschweig, 1961). I give the passage here in my translation:

> As I have already said elsewhere, I find it idle to speculate which came first, the idea or the experiment. I do hope that nobody is of the opinion any more that theories are derived as binding logical conclusions from protocol books, an opinion which was very much fashionable in my student days.

That is—Newton, no less than Descartes, had preconceived ideas. The fundamental difference lies in the sort of question they asked. Newton symbolizes that kind of scientific temper which chooses a well-formulated scientific problem and tries to solve it; probably, the method by which it is achieved is by trying out preconceived ideas, or in Popper's terms by 'conjectures and refutations'. On the other hand the Cartesians and the Leibnizians ask very general, metaphysical questions, to which the answer must necessarily be an *a priori* principle. Both kinds of approach had great representatives, scientists and philosophers as well; progress in science is to some extent a ball game between the two teams. Descartes's vague conservation principle gave birth to the scalar, analytical approach to mechanics; Newton's physics built the vectorial force-centred mechanics. Electrodynamics owes as much to Maxwell as it does to Faraday. Helmholtz, in my opinion, had a scientific temper which is described here as 'Cartesian–Leibnizian'.

32. A. Einstein in the Herbert Spencer Lecture, delivered at Oxford, 10 June 1933. Printed in Oxford at the Clarendon Press, 1933.

II

THE TRADITION IN MECHANICS

Words: 'energia'

It all started with Aristotle, whose use of the word profoundly marked European philosophy and science in all stages of their development. The meaning of the term can be explained only together with its famous counterpart 'potentiality'. Consequently, before examining the purely physical developments, a brief historical survey of these two terms is appropriate.

This pair of opposites underlies the most important problem in Aristotelian physics, namely that of motion. Matter is the potential thing actualized by the 'energy' of form; but matter and form are inseparable, inasmuch as the actual is itself potential having reached completion. The example mostly used by Aristotle and by his followers in the centuries after him is the excess stone which a sculptor chips away, when making the form appear, i.e. potentiality brought into actual existence. The concept is intimately connected with the end-in-itself, i.e. the 'entelechy' of the thing.

Werner Jäger emphasizes the mind-context of 'potentiality' when explaining that in Aristotelian psychology, the mind activates and actualizes its own possibility to the full; actualization (*energia*) occurring by the 'hard work' of thinking (*ergon*), a concept associated with the image of seed-turning-into-plant. Whether the basic context was biological (which is the earlier Jäger theory) and is only then generalized into a metaphysical principle—or rather logical-ontological (the second, later, Jäger theory)—which found its application in the sciences (in our sense) is difficult to decide; there is no single agreed conclusion among the authorities. The choice in this matter would largely dictate our interpretation of Aristotelian physics. For our purposes it will suffice to point out, that the seed-into-plant image accompanying the concept of potentiality is the one most acceptable, even to twentieth-century readers. Qualitatively, it does not differ from the marble-to-statue example. Both are misleading, insofar as they are what we retain from Aristotle's much richer concept. The most difficult, and to us strangest, usage, is Aristotle's definition of motion.

The fulfilment of what exists potentially, insofar as it exists potentially, is motion.[1]

In applying this definition to the case of a stone on the ground, one can either say that this stone has now the nature of act, having reached its natural place, or that it has the nature of potency with respect to the violent force which will throw it upwards after which it will return to its original position and have the nature of act.

This interpretation is in agreement, it seems to me, with what Aristotle says further:

The same thing, if it is of a certain kind, can be both potential and fully real, not indeed at the same time or not in the same respect . . .[2]

The same interpretation is implied by what Aquinas notes:

He says, therefore, that it was necessary to add 'insofar as it is such' because that which is in potency is also something in act. . . . It is clear that the same subject is in potency to contraries as humour or blood is the same subject which is potentially related to health and sickness. And it is also clear that to be in potency to health and to be in potency to sickness are two different things.[3]

The third neo-Platonic philosopher, Plotinus (A.D. 205–70), who criticized Aristotle's physics on many essential points, in his *Enneads* deals with the problem of 'potentiality v. actuality' at length. He begins with an Aristotelian distinction between the two concepts and then goes on to clarify the concept of potentiality:

We cannot think of potentiality as standing by itself; there can be no potentiality apart from something which a given thing may be or become. Thus bronze is the potentiality of a statue; but if nothing could be made out of the bronze, nothing wrought upon it, if it could never be anything as a future to what it has been, if it rejected all change, it would be bronze and nothing else.[4]

Sambursky explains what is happening here:

In the later post-Aristotelian period an increasing need was felt to express the necessity for such capacity within the frame of scientific terminology. Potentiality is only a necessary condition for actuality but it need not be a sufficient one. . . . The technical term signifying the sufficient condition for actualization was *epitedeiotes*, meaning fitness, appropriateness or suitability, and it came into use as a definite scientific concept in the second century A.D.[5]

Having this in mind, it is easier to understand Plotinus' conclusion that 'potentiality may be thought of as a Substratum to states and shapes and forms which are to be received . . .'. Here begins the evolution of the concept of potentiality which is analogous to the peeling of an onion. Generation after generation the concept lost one or more of its layers of meaning. In Plotinus the concept becomes less relative and thus less flexible, and seemingly clearer; however, one wonders whether one could explain Aristotelian motion (which Plotinus did not replace by something better) by using this modified version of the concept of potentiality.

The great Aristotelian interpreters of Nature, Albertus Magnus and St Thomas Aquinas, accepted in principle that the most original Aristotelian contribution to physical science consisted in the discovery of pure potentiality as a reality. But being Christian theologians they took one further step, which to the author seems to have been decisive in the evolution of the concept:

Albertus Magnus and St Thomas never confused Aristotelian physical theory with metaphysics.[6]

This separation, which had been kept up by the Scholastic philosophers, made it feasible that when Galileo rejected Aristotelian physics, the concepts of potentiality and actuality disappeared for some time from pure physical theory.[7]

Words: 'energy'

From now on the history of 'energy' and the history of 'potentiality' part. The intriguing combination emerging now is the 'force-energy' complex, and this is the topic which interests us; the reason I went so far back into the history of the other pair of concepts is that they left an indelible mark on philosophy, and even when banned from physics they made a rear entrance via philosophy. The best proof for this is Leibniz who, as is well known, was the modern thinker most familiar with the Scholastics. Moreover, his chief aim was to try to rescue what he considered good in that school, and to incorporate the remains in his philosophical system. I shall return to Leibniz's dynamics in more detail elsewhere, but at this point his use of the word 'energy' rates emphasis. Leibniz uses the word fairly often, not in connection with his conception of force, which he termed '*vis viva*', but in completely different contexts. (The following also constitutes a glaring example of how misleading careless translations can be!) In his 'On Nature itself, or on the Inherent Force and Actions of Created Things' (1698)[8] Leibniz poses two questions to M. Sturm, with whom he is engaged in a discussion, the second being 'whether there is any energy in created things?'[9] In the original Latin, one finds that Leibniz was so careful to make his intention clear that he even used the Greek ἐνεργεια; thus clearly indicating that what he had in mind was the Aristotelian usage of the term. It is noteworthy that Leibniz was also a linguist, keenly sensitive to the importance of language and the use of terms, and an extremely lucid thinker.[10]

The most persistent usage of the word 'energy' in English, and the less common but no less persistent usage of 'Energie' in German or French, is literary. This will be illustrated from the writings of Humphry Davy in the next chapter. The *Oxford English Dictionary* defines energy as 'force or vigour of expression' and traces it back to 1599. From the 1650s it also has the meaning of 'exercise of power', and this broad usage was retained even after the 1850s, especially by metaphysicians who did not easily abandon their theory of forces. In 1859 Sir William Hamilton, the Scottish philosopher, wrote in a lecture: 'The faculty of which this act of revocation is the energy, I call reproduction.'[11] Mrs Sommerville, who certainly knew physics and was engaged in the raging controversy about conservation ideas, and of 'forces' and 'energies', still spoke in 1849 about the 'disturbing energy of the planets'.

The first modern usage of the word in English was that by Thomas Young in his 1807 *Lectures on Natural Philosophy*.

The term energy may be applied with great propriety to the product of the mass or weight of the body, into the square of the number expressing its velocity. . . . This product has been denominated the living or ascending force . . . some have considered it as the true measure of the quantity of motion; but although it has been very universally rejected, yet the force thus estimated well deserves a distinct denomination.[12]

Not only is this usage clearly confined to mean what we now call 'kinetic energy' but Young's use makes it very clear that he has no conservation ideas in mind at all. This is seen in two ways. First he speaks very clearly of 'forces being consumed' (e.g. 'much more of the force is consumed in producing rotatory motion, than in the second, and that it therefore descends much more slowly'); he also makes it clear that there is no direct connection between work and his 'energy':

The exertion of an animal, the unbending of a bow, and the communication of motion by impulse, are familiar instances of the actions of forces. We must not imagine that the idea of force is naturally connected with that of labour or difficulty; this association is only derived from habit, since our voluntary actions are in general attended with a certain effort, which leaves an impression almost inseparable from that of the force that it calls into action.[13]

Though this will be mentioned again later, it is noteworthy that Young opposes the term 'force of inertia', inertia not being a 'cause of change of motion', and that he reserves the term 'force' for a 'cause of a change of motion' with respect to 'quiescent space'. A digression is inevitable at this point. Young introduced his term of 'energy' very appropriately in a chapter called 'On Collision'. The whole problem of the mechanical conservation of energy, or the principle of conservation of *vis viva*, is intimately connected with the question of elastic and inelastic collision. Our physical hindsight leads us to the conclusion that if already in 1669 Huygens, Wren and Wallis described correctly such collisions (it is usual to assume in books on the history of science that Huygens and Wren dealt with elastic collisions and Wallis with inelastic ones), then they must at least have had clear ideas of kinetic energy. This is a very complicated issue, which we shall not try to disentangle at this stage. Suffice it to say that from careful examination of the sources the author concludes that until 1726 (the date of the second prize essay by the French Académie des Sciences on this problem) some held that 'hard bodies' are inelastic bodies. To illustrate the extent of the confusion on this subject we quote two well-known historians of science. Dijksterhuis in his *The Mechanization of the World Picture* says:

For the special case of perfectly hard bodies (by which it is also meant, as in Huygens that they are perfectly elastic) Descartes frames seven rules of impact.[14]

On the other hand in Herivel's *The Background to Newton's Principia* we find:

Notice that for Newton absolutely solid meant perfectly elastic, whereas for Descartes in his *Principia Philosophiae* it implied perfectly inelastic.[15]

Needless to say, the clarification of what kind of collision they are talking about precedes the attribution of any conservation laws to them.[16] The historical reason for connecting the idea of *vis viva* with the term 'energy' is very difficult to ascertain, but let me suggest a solution. Huygens repeatedly mentions 'efficacy or *vis viva*' as two synonyms. Now 'efficacio' is a term used also by Newton, for something very vague like 'power'. In Definition VIII of Book I of the *Principia*, Newton writes: 'vim acceleratricem ad locum corporis, tanquam efficaciam quandam, de centro per loca singula in circuitu diffusam, ad movenda corpora quae in ipsis sunt'. This 'tanquam efficaciam quandam' was translated by Motte as 'a certain power or energy'. Interestingly, in Cajori's revision the term 'energy' was deleted, probably to facilitate understanding.

There were also suggestions to apply the term 'energy' to what we call 'momentum'. This usage very probably goes back to Colin Maclaurin. Why this excellent mathematician, who seems to have understood the *Principia* so clearly on so many topics, confused these concepts is not clear; but it seems not unnatural that the *Edinburgh Review* in 1808 (xii, p. 130) suggested that: 'This modification of power (that of a moving body proportional to the quantity of matter multiplied into the velocity) might be called "Energy".'

In the entry 'Energie' in the *Grande Encyclopédie*, written by d'Alembert, both 'energie' and 'force' are mentioned only in the literary sense for, otherwise, according to him, the difference is clear.

Nous ne considerons ici ces mots qu'en tant qu'ils s'appliquent au discours; car, dans d'autres cas leur différence saut aux yeux.[17]

This brings us to the term 'force', or its Latin form '*vis*'. Here dictionary treatment will not help us much, and we would have to follow the developments of the concept of 'force' from Newton, Leibniz, Euler, d'Alembert and Lagrange on the one hand, to the vague philosophical 'forces of nature' on the other. This is a task for a special study, and has been recently developed in great detail by many historians.

Let me refer here to Max Jammer's *Concepts of Force*, which traces very well the emancipation of physics from the Scholastics' misuse of Aristotelian physics, through the fundamental concepts in the seventeenth century. As to the development of Newton's thought on dynamics, the above quoted book by Professor John Herivel is the most recent one, while the definitive treatment of Newton's concept of force will soon appear in a long paper by Professor I. Bernard Cohen: 'Newton's Second Law and the Concept of Force in the *Principia*'. Professor Richard Westfall is engaged in writing a monograph on the seventeenth-century concepts of force. While I shall treat briefly the force-concepts of d'Alembert and Lagrange, I have to skip Leibniz's concept of '*vis*' here. In reality Leibniz's concept of '*vis*' is much more a forerunner of the Naturphilosophie's concept of force, than of our concept of energy.

If we wish to understand the Leibnizian concept of force, we should look into all the works which Leibniz considered relevant thereto—and there are very few of his works which he did not consider relevant to his physics.

The search for a unitary principle goes through the work of Leibniz like a 'Leitmotif'. In his early dynamical works[18] he spoke of such a fundamental principle as conservation of force. What is this force to which the Leibnizian universe can be reduced? It is nothing like the Newtonian force-vector. It is certainly not energy as some modern commentators would like to translate it. What Leibniz actually tells us himself is that force has *an effect* $m\mathbf{v}^2$, or $m\mathbf{v}$, or the height reached by a body thrown upwards, as the case may be, while in his later works he says that it is 'a metaphysical entity', 'the essence of matter' or 'the main attribute of a monad'.

The monad serves as a final generalization of his concept of force, now uniting in it not only all the physical effects of this fundamental entity which are conserved in nature but also the physical and the spiritual: mind and matter. To us, scientifically-minded, logically-trained moderns all this sounds very confused; but the importance of vague concepts like Leibniz's force or monad for creating science is enormous.[19] Most scientific concepts at early stages of their evolution as well as theories in which they occur defy any attempt to decide whether they are distinct or indistinct, whether they belong to science or to metaphysics. It seems to me that historical research on the evolution, importance and influence of such concepts and theories can be useful only if systematically pursued in every work of the same author, or again if we choose a limited period during which we trace these developments in the work of all important authors.

The 'perpetuum mobile' *issue*

The attempts continued for centuries to create power out of nothing have slowly abated. This was the result of the new realization, an inductive conclusion, that man could not construct a perpetual motion machine. Nevertheless, this growing conviction was not shared by all. The great Dutch scientist 's Gravesande writes:

Les Mathematiciens, & les Physiciens, sont partagés sur la Force du choc. Les uns croient, & c'est le sentiment le plus ordinaire, que les Forces des differents chocs d'un même corps sont entre elles, comme les vitesses de ces corps. Les autres, au contraire, soutiennent que ces même Forces sont entre elles comme les carrés des vitesse. Tous conviennent que la Force du choc est proportionelle à la masse; c'est pourquoi les premiers multiplient la masse par la vitesse, pour avoir la Force du choc; les autres multiplient la masse par le carré de la vitesse, pour determiner cette même Force.

Je n'examinerai pas ici le quel de ces deux principes se conforme à l'experience: je me propose de faire voir: (1) Qu'en admettant le premier il faut admettre la possibilité du Mouvement perpetuel, dans les Machines qui auront pour principe de leur mouvement le choc des corps. (2) Qu'en admettant le second principe, l'impossibilité du Mouvement perpetuel n'a pas encore été demonstrée dans tous les cas possibles. Et (3) enfin, je tâcherai de faire voir que les loix de la nature ne nous sont pas assez connues pour en tirer une conclusion générale que le Mouvement perpétuel est contraire à ces loix.[20]

There are many complex problems hinted at in this passage; 's Gravesande did important experimental work, and defended the Leibnizian conception of the conservation of *vis viva*.[21] But if he accepted the conservation of *vis viva*, how could he possibly have advocated that perpetual motion is not contrary to the laws of nature? Alternatively, these conservation ideas could have been so strange to him that he actually accepted that 'force' should indeed be measured by *vis viva* without taking it to be a conserved entity. But this would imply that the concept was so new to him that he misunderstood Leibniz. For, as we have seen, Leibniz did not have an idea of 'energy', but he certainly had an idea of conservation for those forces which he chose to call by the name of live forces.

Others had read the Leibnizian message correctly. De Miran, in his 1728 memoire, 'Dissertation sur l'estimation et la mesure des forces motrices des corps', wrote:

La force vive dit-il [speaking of the antagonists of the *vis viva*] est celle qui réside dans un corps lorsqu'il est dans un mouvement actuel. C'est cette force qu'on fait proportionelle au carré de la vitesse. Mais il faut qu'elle soit actuellement exercée dans la communication du mouvement et pendant un temps fini. Pour se manifester et pour se montrer proportionelle au carré de la vitesse, elle devient par là toute differente de la force morte et, en un sens, de celle que nous avons considerée dans la choc du corps infiniment dur. Elle ne peut ni naître ni périr en un instant, il faut plus ou moins de temps pour la produire ou pour la détruire.[22]

While this theoretical discussion was going on, new devices and machines were presented to the Académie des Sciences proving the futility of the quest. Finally, in 1775 the Académie des Sciences lost patience with the problem and pronounced it settled. Before quoting that decision, let me once more take up the claim, mentioned in the Introduction, that this realization of the impossibility of a Perpetual motion machine is the real source of the principle of conservation of energy. We have seen that serious physicists like 's Gravesande did not even connect the two ideas! Those who tried experimentally to build a machine had little contact with theoretical developments: some of them inductively reached the conclusion that, they having failed, such a machine cannot be constructed, while others drew the conclusion that having failed they had to continue trying until success would come. On the other hand those who accepted the impossibility, did so on the basis of their metaphysical commitments to some kind of conservation principle, such as 'nothing can be made of nothing' or that 'cause must equal the effect'. After the final refusal by the Académie des Sciences to consider any more attempts, nothing occurred for almost three-quarters of a century to bring the principle of conservation of energy nearer. Trying to cut the long and tedious list of the discussants, even the famous Johannes Bernoulli was not quoted, although he did advocate the possibility of the perpetual motion machine. The point I tried to make was, that when looking into these causal relations of development, the realization of the impossibility of a *perpetuum mobile* had little to do with the establishment of the principle of conservation of energy. In other words, the fact that the principle of conservation does imply the impossibility of a perpetual motion machine, is not a sufficient condition for it but only a necessary one. This conclusion will be further strengthened when we examine the contributions of d'Alembert and Lagrange more carefully. But first, for the sake of completeness, here is the text of the Académie Royale des Sciences in full:

The resolution of the Royal Academy of Sciences in Paris not to entertain communications relating to Perpetual Motion, was passed in 1775 and reads as follows:

This year the Academy has passed the resolution not to examine any solution of problems on the following subjects:

The duplication of the cube, the trisection of the angle, the quadrature of the circle, or any machine announced as showing perpetual motion.

We believe ourselves bound to account for the motives which have led to this determination.

*   *   *

The construction of a perpetual motion machine is absolutely impossible. If even friction and resistance from the middle did not eventually destroy the effect of the first motive power, that power cannot produce an effect equal to its cause; if, then, it is desired that the effect of a complete power should act continually, the effect must be infinitely small in a given time. If the friction and resistance be subtracted, the first motion given to a body will always continue; but it will not act in regard to other bodies, and the only perpetual motion possible in this hypothesis (which could not exist in nature) would be absolutely useless in carrying out the object proposed by the constructors of these perpetual motion machines. The drawback to these researches is their being exceedingly expensive, and has ruined more than one family; often mechanics, who could have rendered great services to the public, have wasted their means, time and genius.

Such are the principal motives that have dictated the determination of the Academy. In stating that they will not occupy themselves any longer with these subjects, they only declare their opinion of the complete uselessness of the labour of those who so occupy themselves. It has often been said, that in seeking to solve chimerical problems, many useful truths have been found; an opinion which originated in a time when the proper method of discovering the truth was unknown, which in the present day is well known. It is more than probable that the right manner of discovering these truths is to search for them. But the quadrature of the circle is the only rejected problem of the Academy which could give rise to any useful research; and, if a geometrician should find it out, the determination of the Academy would only enhance his merit, as it would show the opinion that geometricians have of the difficulty, not to say insolubility, of the problem.[23]

Rational mechanics in general

In the first half of the eighteenth century, the development of mechanics was in the hands of the mathematicians. The landmarks in this development are Euler's *Mechanica* in 1736, d'Alembert's *Traité de Dynamique* in 1743, and Lagrange's *Mécanique Analytique* in 1788. By that time the concepts in which mechanics was analysed, had been hammered out, except the concept of 'force' which was still in a state of flux, and the concept of 'energy' which had not yet been born. There were also two conservation principles: momentum ($m\mathbf{v}$) was believed to be conserved under all conditions (after the clarifications of Huygens and Leibniz that it was the vectorial quantity of momentum which was conserved), and a fairly clear notion that the scalar quantity $m\mathbf{v}^2$ was conserved at least in elastic collisions. Both were considered forces of some sort; the first conservation law did not enable the calculation of velocities after collision (because of its vector character), while for the second law it was not clear what happens in inelastic collisions. Every investigator performed different experiments, and in view of the special case of his experiments gave different names to the entities involved; the very fact that none of them realized that all these experiments on falling bodies, compressed springs, clay cylinders, colliding balls of glass, clay, wool, etc., were obeying the same laws shows how superficial their conservation laws were. Lagrange, who was not much troubled by terminology or even conceptual difficulties, talking about Galileo said: 'Galileo entend par moment d'un poids ou d'une puissance appliquée à une machine, l'effort, l'action, l'énergie, l'impetus de cette puissance pour mouvoir la machine.'[24] He could have added several more terms like *vis viva, vis mortua, vis potentia*, and *vitesse virtuelle* and for each could have found several of the greatest mathematicians and physicists of the age who would all have sworn that theirs was the correct term. I shall not go into these problems in detail; let me remark only that the last one—'*vitesse virtuelle*'—is a term introduced by Jean Bernouilli in a letter to Varignon (26 January 1717); he then called the virtual velocity multiplied by the force 'energy'. The law that has been accepted by the time of d'Alembert's treatise, and later put into the centre of mechanics by Lagrange, was the following (I quote from Erwin Hiebert's *Historical Roots of the Principle of Conservation of Energy*):

The formulation of the conservation law for mechanical energy had its scientific roots in at least three areas of theoretical mechanics. (1) The principle of conservation of mechanical (or virtual) work—that in every transfer of one form of potential energy into another form the total energy remains unchanged. (2) The principle of conservation of *vis viva* (or quantity of motion as it was first designated, but without precision)—that in every transfer of one form of kinetic energy into another form (for example, in an elastic collision), the total energy remains unchanged. (3) The principle of conservation of (1) and (2) taken conjointly (the modern concept of conservation of energy)—that in every transfer of potential energy into kinetic energy and vice versa the total energy remains unchanged.[25]

This is written in modern terms, and summarizes admirably the limited conservation principle as adopted by d'Alembert and others, as applying to ideal mechanical systems, but it is not the 'modern concept of conservation of energy'—it is pure hindsight. To be able to say that this is a special case of a more general law to which it became generalized, presupposes conceptually the knowledge of that general law. The same physicists who adopted this elegant mathematical formulation, if they cared at all (for example Lagrange did not—he had very little to do with physical 'reality') were completely satisfied to accept, that in some non-idealized situations '*vis viva*' or '*vis potentia*' get lost completely. If Leibnizians, they talked about a vague conservation of forces, thus preparing the ground for the next century's 'Naturphilosophie', and then acknowledged that even in inelastic collisions nothing is lost completely for 'cause must equal effect'.

The rational mechanics that developed now is, in the words of Professor Truesdell,

neither experimental nor philosophical; it is mathematical; it is a history of special problems, concrete examples for the solution of which new principles and methods had to be created. . . . Rational mechanics was a science of experience, but no more than geometry was it experimental. . . . Experiment and theory result from different kinds of reaction to experience. If, ideally, they should complement and check one another (and even today, with all our superior knowledge not only of facts but also of scientific methods, it is difficult to relate them), why should it have been easier 300 years ago? It was not. A factual view of the history of mechanics must concede that rational mechanics, both arising from human beings' intelligent reaction to mechanical experience, grew up separately.[26]

Euler

Euler's *Mechanica* is an early work, very much influenced by Newton and carrying out the Newtonian programme in mathematical language. He accepts the Newtonian concept of force, and wherever that is not crystal clear, Euler tries to make it so. For him (in this work!) power (*potentia*) or force (*vis*) is characterized by the modification of the motion of a particle that is produced by it. A power is directional:

Potentia est vis corpus vel ex quiete in motum perducens, vel motum ejus alterans. Directio potentiae est linea recta secundum quam ea corpus movere conatur.[27]

And also the force of inertia is a force like any other:

Vis inertiae est illa in omnibus corporibus facultas vel in quieto permanendi vel motum uniformiter in directum continuendi.[28]

That he built his concept of force primarily on statics emphasizes how clearly his '*vis inertiae*' was defined. At this stage he also accepted the Newtonian 'absolute' versus 'relative' qualities.

This Newtonian influence is expressed also in the structure of the two volumes of the *Mechanica* with its Definitions, Demonstrations of Theorems and Scholia. His usage of a 'scholium' is exactly as that of Newton; and here again he makes it even clearer: if in Newton it can be asked what is the exact difference between a 'Theorem' and a 'Problem', here it is clear that both are 'Propositions'; a 'Theorem' has a 'Demonstration', while a 'Problem' has a 'Solution'; also the wording is somewhat different, a 'Problem' including the expression 'determinare'. My emphasis as to the date was introduced in view of the fact that Euler diverged much from his Newtonianism in later years, and was more and more influenced by that same Leibnizian philosophy which he so much attacked; with this process his concept of force changed greatly. Nevertheless, the early stage of his work is very interesting for us, not only because it is great work, but because it was this work which had the decisive influence on d'Alembert.[29]

D'Alembert

On many levels d'Alembert is the most fascinating among the French mathematicians. His literary activity, his prominent influence on the French intelligentsia, his scientific work, all add up to a most worthwhile study. But this is not a study of d'Alembert and we must limit ourselves to the few relevant points. D'Alembert has often been accused of being a sloppy mathematician; he would have certainly taken offence because he prided himself mainly on his mathematical ability. He did not realize that he did important conceptual work in mechanics—perhaps the only one to have done so among the great creators of rational mechanics. His conceptual contribution has to do with his concept of force, and his formulation of mechanics as a mediator between the Newtonian, strictly vectorial, formulation and the Lagrangian, strictly scalar formulation. He first published his *Traité de Dynamique* in 1743, and a new edition appeared with many additions in 1758. The 'Discours Préliminaire' is both a detailed programme for mechanics, and a chapter in the philosophy of science. On Cartesian lines, he aims to establish the science of mechanics on simple and clear ideas, the principles of mechanics, which, in his opinion are necessary truths. But these principles are not those of Descartes (as regards extension, and the clear and distinct idea of motion) nor those of Euler (as regards the importance of impenetrability; an idea that does not yet play a dominant rôle in his *Mechanica* but returns later in his much abused philosophical writings as prior in importance even to Newton's laws of motion) but three distinct mathematical principles:

Ce paradoxe ne paraîtra point tel à ceux qui sont étudié ces sciences en philosophes; les notions les plus abstraites, celles que le commun des hommes regarde comme les plus inaccessibles, sont souvent celles qui portent avec elles une plus grande lumière; . . . l'impenetrabilité, ajoutée à l'idée de l'étendue, semble ne nous offrir qu'un mystère de plus; la nature du mouvement est une enigme pour les philosophes; le principe metaphysique des lois de la percussion ne leur est pas moins caché.[30]

His work will be based on three distinct principles other than these: the force of inertia, the principle of compound motion, and the principle of equilibrium. The passage where this is most clearly expressed is the following:

The principle of equilibrium together with the principles of the force of inertia and of compound motion, therefore leads us to the solution of all problems which concern the motion of a body in so far as it can be stopped by an impenetrable and

immovable obstacle—that is, in general by another body to which it must necessarily impart motion in order to keep at least a part of its own. From these principles
together can easily be deduced the laws of the motion of bodies that collide in any
manner whatever, or which affect each other by means of some body placed in
between them and to which they are attached.[31]

Dugas also claims that 'Lagrange said, and is often repeated, that
d'Alembert had reduced dynamics to statics by means of his principle'.
Now, it is true that Lagrange deals with this problem in the historical
chapter in his *Mécanique Analytique*, and it is also true that these historical
notes were very inaccurate, misleading and biased, and that they formed
the foundation of many more misleading and biased histories to come—
one good example is Mach's *Mechanics*; but however biased Lagrange may
have been he is certainly a good authority on his own views about somebody else. He writes:

Cette manière de rappeler les lois de la Dynamique à celles de la Statique est la
vérité moins directe que celle qui resulte du principe de d'Alembert, mais elle
offre plus de simplicité dans les applications; elle revient à celle d'Herman et
d'Euler qui l'a employée dans la solution de beaucoup des problemes de Mécanique, et on la trouve dans quelques Traités de Mécanique sous le nom de Principe
de d'Alembert.[32]

What d'Alembert did will be summarized in short in modern terminology. We start with the fundamental Newtonian law of motion (as
actually first formulated by Euler)

$$m\mathbf{a} = \mathbf{F},$$

after that we rewrite this equation in the form

$$\mathbf{F} - m\mathbf{a} = 0.$$

Now we can define a vector $\mathbf{I}$ (inertia) as

$$\mathbf{I} = -m\mathbf{a}.$$

This vector considered as a force is called 'the force of inertia' and
satisfies

$$\mathbf{F} + \mathbf{I} = 0.$$

Now what has been gained? This is a most difficult question that any
first year university student poses to himself or to his teacher. The answer
is, that what has been gained is not so much mathematical but a conceptually new principle. The Eulerian formulation ($\mathbf{F} = m\mathbf{a}$) was already a great
step in the direction of a clear mathematical formulation which stifles
philosophical debates, as to what Newton could have meant by '*Vis
inertiae*', or whether his 'force' is an $\mathbf{F}$ or $F dt$; but it was still not sufficiently

clear conceptually whether those two entities—external forces which cause the change of motion, and those curious *m***a**'s or d(*m***v**)'s—are entities physically commensurable or not. Are they the same kind of force? Euler made it very clear that for him forces were primary entities, and that force of inertia was a force like any other. This, d'Alembert learned from him, or read into him. In any case d'Alembert, presenting his mechanics in a vectorial formulation and combining the Newtonian heritage with Euler's *Mechanica*, was the right man to conceive of the force of inertia as a force which can be added to and subtracted from the vector sum of all other forces acting on the body. In all probability, he was the last man who could have invented d'Alembert's principle. For the further conclusions from this very principle were such that the scalar treatment was called forth and readily developed by Lagrange into the crowning masterpiece of rational mechanics. This combination of Newtonian–Eulerian influence made the principle conceptually possible. And finally, only one with so little feeling for an experimental situation (I do not mean the fact that he did not actually perform experiments but rather his lack of an interest in phenomenological theory as against the purely theoretical) so easily could have agreed to considering the force of inertia as just another force; d'Alembert was not committed either to Descartes's metaphysical presuppositions or to Newton's verbal-philosophical formulations, which were enough to discourage any mechanician from the step taken by d'Alembert. Descartes did not deal with the concept of 'force' in these terms, while Newton never really expressed clearly as to whether to consider the 'force of inertia' just another force. If we study his formulations and definitions we will tend to think that he did not. Again, the reader is referred to the extensive treatment of this topic in I. Bernard Cohen's paper, mentioned above.

This interpretation is not challenged by d'Alembert's alleged attempt to ban the concept of force from mechanics. Here again is a confusion of words; d'Alembert did not attempt to ban the Newtonian concept of force—his very principle is based on it. Though d'Alembert did not attribute to any physical concept such primary importance as, for example, Euler had done, the 'force' that he tried to eliminate—at least from the central place it had begun to occupy—was the *'force vive'*! D'Alembert stresses the three different principles (the conservation of *vis viva* is not one of them) and then reviews that famous controversy:

Si les principes de la force d'inertie, du mouvement composé et de l'équilibre sont essentiellement différents l'un de l'autre, comme on ne peut s'empêcher d'en convenir; et si d'un autre côté ces trois principes suffisent à la Mécanique, c'est avoir reduit cette Science au plus petit nombre de principes possibles. . . . Tout ce que nous voyons bien distinctement dans le mouvement d'un corps, c'est qu'il parcourt un certain espace et qu'il emploie un certain temps à la parcourir. . . . J'aie pour ainsi dire, détourné la vue de dessus les causes motrices, pour n'envisager uniquement que le mouvement qu'elles produisent; que j'aie entièrement proscrit les forces inhérentes au corps en mouvement, êtres obscures et metaphysiques, qui ne sont capables de rependre les ténèbres sur une Science claire par elle même.[33]

He rejects the 'forces inhérentes au corps en mouvement', and for that reason 'j'ai cru ne devoir point entrer dans l'examen de la fameuse question des *forces vives*'. It was not an accident that d'Alembert emphasized that the forces he wants to reduce in importance for metaphysical reasons are the forces which inhere in the body while in motion (that is the 'force' called $m\mathbf{v}$ and the 'force' called $m\mathbf{v}^2$). For he repeats it again, very carefully formulating:

Quand on parle de la force des corps en mouvement, ou l'on n'attache point d'idée nette au mot qu'on prononce, ou l'on ne peut entendre par la, en general, que la propriété qu'ont les corps qui se meuvent de vaincre les obstacles qu'ils rencontrent ou de leur résister.[34]

In other words just 'force' for him is one of those two debated quantities, and not the Newtonian 'force', and this debate does not seem to him worthwhile of solution—better to circumvent it. It is naturally true, as Thomas Hankins pointed out,[35] that d'Alembert's explanation of the problem was merely one of words, and also, that finally d'Alembert retained the concept of momentum. But this was not because it was a conserved entity unlike the *vis viva*, but because it was a mathematically clear and distinct entity (in its natural vector form) and thus he could work with it. Conservation ideas are not central to d'Alembert's conceptual framework.

Applying the reduction of dynamics to statics has not yet led to the solution of dynamical problems by statical methods. The resulting equations are differential and have to be solved. The only achievement was the deduction of these differential equations by statical methods. The equations are now of equilibria and here the principle of virtual velocities applies. Accordingly, d'Alembert's principle in a modern formulation reads:

The total virtual work of the effective forces is zero for all reversible variation which satisfy the given kinematical conditions.[36]

Lagrange

Not much work has been done on Lagrange, and all that will be said here is tentative. Apparently Lagrange, whose work was a direct continuation of the work of d'Alembert, was neither interested in nor bothered by philosophical problems. He was a mathematician of the highest rank, with mathematical aims, imbibed with the positivistic spirit of what very soon became the spirit of the École Polytechnique. He gave perfunctory definitions of 'force' and 'power' not really caring what was implied by them, simply trying to write his *Mécanique Analytique* on the model of previous works, so that definitions had to be given. But in reality, forces in the vectorial sense play small importance in his formulation. One could say that the whole work is written in scalar language. Lagrange successfully derived his famous equations from the so-called Newton's laws (in Euler's formulation), and this deduction shows in a very interesting way how the transformation from vector to scalar language took place. In the more elegant modern treatments, where Lagrange's equations are usually deduced from Hamilton's principle—thus exactly reversing the historical order—this transformation is being lost sight of, because the conceptual basis of Hamilton's principle is richer and completely different. The advantages of the Lagrangian–analytical treatment is admirably summarized by Lanczos in the above-quoted book:

The analytical approach to the problem of motion is quite different. The particle is no longer an isolated unit but part of a 'system'. A 'mechanical system' signifies an assembly of particles which interact with each other. The single particle has no significance; it is the system as a whole which counts. For example, in the planetary problem one may be interested in the motion of one particular planet. Yet the problem is unsolvable in this restricted form. The force acting on that planet has its source principally in the sun, but to a smaller extent also in the other planets, and cannot be given without knowing the motion of the other members of the system as well. And thus it is reasonable to consider the dynamical problem of the entire system, without breaking it into parts.

But even more decisive is the advantage of a unified treatment of force-analysis. In the vectorial treatment each point requires special attention and the force acting has to be determined independently for each particle. In the analytical treatment it is enough to know one single function, depending on the positions of the moving particles; this 'work function' contains implicitly all the forces acting on the particles of the system. They can be obtained from that function by mere differentiation.

Another fundamental difference between the two methods concerns the matter of 'auxiliary conditions'. It frequently happens that certain kinematical conditions exist between the particles of a moving system which can be stated *a priori*. For example, the particles of a solid body may move as if the body were 'rigid', which means that the distance between any two points cannot change. Such kinematical conditions do not actually exist on *a priori* grounds. They are maintained by strong forces. It is of great advantage, however, that the analytical treatment does not require the knowledge of these forces, but can take the given kinematical conditions for granted. We can develop the dynamical equations of a rigid body without knowing what forces produce the rigidity of the body. Similarly we need not know in detail what forces act between the particles of a fluid. It is enough to know the empirical fact that a fluid opposes by very strong forces any change in its volume, while the forces which oppose a change in shape of the fluid without changing the volume are slight. Hence, we can discard the unknown inner forces of a fluid and replace them by the kinematical conditions that during the motion of a fluid the volume of any portion must be preserved. If one considers how much simpler such an *a priori* kinematical condition is than a detailed knowledge of the forces which are required to maintain that condition, the great superiority of the analytical treatment over the vectorial treatment becomes apparent.

However, more fundamental than all the previous features is the unifying principle in which the analytical approach culminates. The equations of motion of a complicated mechanical system form a large number—even an infinite number—of separate differential equations. The variational principles of analytical mechanics discover the unifying basis from which all these equations follow. There is a principle behind all these equations which expresses the meaning of the entire set. Given one fundamental quantity, 'action', the principle that this action be stationary leads to the entire set of differential equations. Moreover, the statement of this principle is independent of any special system of coordinates. Hence, the analytical equations of motion are also invariant with respect to any coordinate transformations.

What Lagrange did, again in modern terms first, was to arrive at a set of equations:

$$\frac{\mathrm{d}}{\mathrm{d}t}\left(\frac{\partial L}{\partial \dot{q}_f}\right) - \frac{\partial L}{\partial q_f} = 0$$

where the q's denote generalized coordinates, L is a Lagrangian function $(L = T - V)$, T is what we call kinetic energy, and V is what we call potential energy, which Lagrange called the potential function. There are many remarkable properties of these equations, like their invariance with respect to arbitrary point transformations.[37] But what is more important for us is that they make use of a single scalar function. In modern terms, Lagrange's equations are a differential energy principle. This scalar L determines the whole dynamics of the problem for which the equations were set up.

In Lagrange's own terms, what he did looked as follows: (I give the summary of Dugas, who is a great admirer of Lagrange, and on this topic clearer than any other description which has come to my notice.)

Lagrange was able to put the equations of dynamics into a very general and valuable form which has now become classical.

For each element, of mass m, of a system, Lagrange defines 'the forces parallel to the axes of coordinates which are used, directly, to move it', to be

$$m\frac{\mathrm{d}^2 x}{\mathrm{d}t^2} \qquad m\frac{\mathrm{d}^2 y}{\mathrm{d}t^2} \qquad m\frac{\mathrm{d}^2 z}{\mathrm{d}t^2}.$$

He regards each element of the system as acted upon by similar forces, and concludes that the sum of the moments of these forces must always equal the sum of the given accelerating forces which act on each element. Thus he writes

$$\mathrm{Sm}\!\left(\frac{\mathrm{d}^2 x}{\mathrm{d}t^2}\,\delta x + \frac{\mathrm{d}^2 y}{\mathrm{d}t^2}\,\delta y + \frac{\mathrm{d}^2 z}{\mathrm{d}t^2}\,\delta z\right) + \mathrm{Sm}(P\delta p + Q\delta q + R\delta r + \ldots) = 0$$

the given forces P, Q, R, . . . being supposed to act on each element along the lines p, q, r, . . .

Lagrange transforms the first sum by using the identity

$$\mathrm{d}^2 x \delta x + \mathrm{d}^2 y \delta y + \mathrm{d}^2 z \delta z = \mathrm{d}(\mathrm{d}x\delta x + \mathrm{d}y\delta y + \mathrm{d}z\delta z) - \tfrac{1}{2}\delta(\mathrm{d}x^2 + \mathrm{d}y^2 + \mathrm{d}z^2),$$

by a change of variables in which each differential $\mathrm{d}x$, $\mathrm{d}y$, $\mathrm{d}z$, . . . is expressed as a linear function of the differentials $\mathrm{d}\xi$, $\mathrm{d}\psi$, $\mathrm{d}\varphi$ . . .

Lagrange establishes that if Φ is the transform of the quantity

$$\tfrac{1}{2}(\mathrm{d}x^2 + \mathrm{d}y^2 + \mathrm{d}z^2)$$

then the following equation is identically true.

$$\mathrm{d}^2 x \delta x + \mathrm{d}^2 y \delta y + \mathrm{d}^2 z \delta z = \left(-\frac{\partial \Phi}{\partial \xi} + \mathrm{d}\frac{\partial \Phi}{\partial \mathrm{d}\xi}\right)\delta\xi + \left(-\frac{\partial \Phi}{\partial \psi} + \mathrm{d}\frac{\partial \Phi}{\partial \mathrm{d}\psi}\right)\delta\psi + \ldots$$

Lagrange confines himself to forces P, Q, R, . . . for which the quantity

$$P\delta p + Q\delta q + R\delta r + \ldots$$

is integrable, which, he declares, 'is probably true in nature'. This enables him to suppose that

$$\mathrm{Sm}(P\delta p + Q\delta q + R\delta r + \ldots) = \delta\mathrm{Sm}\pi(\xi, \psi, \varphi, \ldots).$$

The general equations of dynamics are then written in the form

$$\Xi\delta\xi + \psi\xi\psi + \ldots = 0$$

by putting

$$\Xi = \mathrm{d}\frac{\partial T}{\partial \mathrm{d}\xi} - \frac{\partial T}{\partial \xi} + \frac{\partial V}{\partial \xi}$$

with

$$T = \tfrac{1}{2}\mathrm{Sm}\!\left(\frac{\mathrm{d}x^2}{\mathrm{d}t^2} + \frac{\mathrm{d}y^2}{\mathrm{d}t^2} + \frac{\mathrm{d}z^2}{\mathrm{d}t^2}\right) \text{ and } V = \mathrm{Sm}\Pi.[38]$$

Lagrange saw immediately that from his equations the principle of conservation of *vis viva* could easily be deduced and he saw the importance of that step. Again quoting Dugas:

Better than d'Alembert had been able to do, Lagrange established that the conservation of living forces is a consequence of the equations of dynamics, as long as the constraints are without friction and independent of time.

For this purpose, Lagrange considers the true motion of the system between the time t and the time $t + \mathrm{d}t$; that is, he substitutes $\mathrm{d}x$, $\mathrm{d}y$, $\mathrm{d}z$, . . . and $\mathrm{d}p$, $\mathrm{d}q$, $\mathrm{d}r$, . . . for δx, δy, δ^2 . . . and δp, δq, δr, . . . in the general formula. This enables him to write

$$\mathrm{Sm}\!\left(\frac{\mathrm{d}x\mathrm{d}^2 x + \mathrm{d}y\mathrm{d}^2 y + \mathrm{d}z\mathrm{d}^2 z}{\mathrm{d}t^2} + P\mathrm{d}p + Q\mathrm{d}q + R\mathrm{d}r + \ldots\right) = 0.$$

If the quantity

$$P\mathrm{d}p + Q\mathrm{d}q + R\mathrm{d}r + \ldots$$

is integrable, then

$$S\left(\frac{\mathrm{d}x^2 + \mathrm{d}y^2 + \mathrm{d}z^2}{2\mathrm{d}t^2} + \Pi\right)m = H.$$

'This equation includes the principle known by the name of the *conservation of living forces*. Indeed, since $\mathrm{d}x^2 + \mathrm{d}y^2 + \mathrm{d}z^2$ is the square of the distance which the body travels in the time $\mathrm{d}t$, then $\dfrac{\mathrm{d}x^2 + \mathrm{d}y^2 + \mathrm{d}z^2}{\mathrm{d}t^2}$ will be the square of the velocity and $m\dfrac{\mathrm{d}x^2 + \mathrm{d}y^2 + \mathrm{d}z^2}{\mathrm{d}t^2}$ will be its living force. Therefore $S\left(\dfrac{\mathrm{d}x^2 + \mathrm{d}y^2 + \mathrm{d}z^2}{\mathrm{d}t^2}\right)m$ will be the living force of the whole system, and it is seen, by means of the equation concerned, that this living force is equal to the quantity $2H - 2\Pi\Pi m$ which only depends on the accelerating forces which act on the bodies and not on their mutual constraints. So that the living force is always the same as that which the bodies would have acquired if they had moved freely, each along the line that it described, under the influence of the same powers.'

Thus Lagrange discovers the same principle as that formulated by Huygens to be a simple corollary of his general equations.[39]

The enormous importance of the Lagrangian formulation slowly dawned on the mathematicians. Many more problems were amenable to this treatment. It was this practical mathematical gain which turned the focus of interest on these new energetic functions, which suddenly appeared in a new light. Still, this was only what we call mechanical energy, and by that we already presuppose the existence of a general concept of energy of which this is a special, though very important, case. I cannot repeat it often enough—this is sheer hindsight. For them it was this old concept of '*vis viva*' and the old concept (though less old than that of the *vis viva*) of 'potential function' which obeyed a conservation law for the special case of no friction and time-independent constraints, and which thus opened a way to the solution of many problems of which mathematicians had already despaired. Those functions do not characterize particles but rather processes. The fact that something is conserved in the system makes the reidentification of the system possible. It had also other far-reaching consequences with which we are not dealing here: the first buds of the concept of field originated in this concept of scalar, conserved, mechanical energy.

Further developments were only partially theoretical; here the applications of mechanics to various mechanical devices and engines becomes more and more prominent, and with it the various concepts of work. Such was the work of Coriolis, Poncelet, Poisson, Lazare Carnot and many others. I shall forgo the list of their contributions. In résumé, they made mechanics a strong theoretical-experimental unit with a clear idea of conservation for those mechanical quantities, under some special conditions. Whether they believed that in cases where there was friction, the mechanical 'energies' got lost or were conserved, was for them not a matter of science, but rather a metaphysical commitment.

Coriolis introduced the factor 1/2 into the *vis viva* in his 'Calcul de l'effet de machines' in 1829. Even before that, Jean Bernoulli developed Leibniz's work in more detail and began to develop a concept of work; in 1826 Poncelet in his 'Cours de mécanique appliquée aux machines' introduced the term 'travail'. This experimental research, or other research which we would nowadays call theoretical engineering, started with Lazare Carnot; in 1783 (before Lagrange's *Mécanique Analytique*!) he published his 'Essai sur les machines en général' and in 1803 his 'Principes généraux de l'équilibre et du mouvement'. Here he introduced new concepts, and very probably indirectly exerted an influence on his son Sadi Carnot, which was to be one of the first steps in connecting mechanical engineering with concepts from the theory of heat.

Before seeing how Helmholtz dealt with the Newton–Euler–d'Alembert–Lagrange heritage, let us check a few claims about some early enunciations of the principle of conservation of energy.

Conservation of energy in general

The central tenet of this study is to show, that in the first and most general formulation of the principle of conservation of energy in Helmholtz's 1847 paper, the tradition in mechanics, the various theories of heat, the physiological thought of those times and some philosophical principles were all necessary ingredients. There have been many histories of the principle, written from many points of view: there is Mach's[40] *History*; Planck's *Das Prinzip der Erhaltung der Energie* in 1887; Hiebert's recent book and many others. I shall not try even to refer to all of them. Rather I would like to point out some of the difficulties with this principle, and deal with two special cases.

Firstly, there are many inherent ambiguities connected not only with the concept of energy, but also with the concept of conservation. This was most clearly stated by Poincaré, and I shall quote him:

Suppose an isolated system formed of a certain number of material points; suppose these points subjected to forces depending only on their relative position and their mutual distances, and independent of their velocities. In virtue of the principle of the conservation of energy, a function of forces must exist.

In this simple case the enunciation of the principle of the conservation of energy is of extreme simplicity. A certain quantity, accessible to experiment, must remain constant. This quantity is the sum of two terms; the first depends only on the position of the material points and is independent of their velocities; the second is proportional to the square of these velocities. This resolution can take place only in a single way.

The first of these terms, which I shall call U, will be the potential energy; the second, which I shall call T, will be the kinetic energy.

It is true that if $T+U$ is a constant, so is any function of $T+U$, $\gamma(T+U)$. But this function $\gamma(T+U)$ will not be the sum of two terms, the one independent of the velocities, the other proportional to the square of these velocities. Among the functions which remain constant there is only one which enjoys this property, that is $T+U$ (or a linear function of $T+U$, which comes to the same thing, since this linear function may always be reduced to $T+U$ by change of unit and of origin). This then is what we shall call energy; the first term we shall call potential energy and the second kinetic energy. The definition of the two sorts of energy can therefore be carried through without any ambiguity.

It is the same with the definition of the masses. Kinetic energy, or *vis viva*, is expressed very simply by the aid of the masses and the relative velocities of all the material points with reference to one of them. These relative velocities are accessible to observation, and, when we know the expression of the kinetic energy as function of these relative velocities, the coefficients of this expression will give us the masses.

Thus, in this simple case, the fundamental ideas may be defined without difficulty. But the difficulties reappear in the most complicated cases and, for instance, if the forces, in lieu of depending only on the distances, depend also on the velocities. For example, Weber supposes the mutual action of two electric molecules to depend not only on their distance but on their velocity and their acceleration. If material points should attract each other according to an analogous law, U would depend on the velocity, and might contain a term proportional to the square of the velocity.

Among the terms proportional to the squares of the velocities, how to distinguish those which come from T or from U? Consequently, how to distinguish the two parts of energy?

But still more; how define energy itself? We no longer have any reason to take as definition $T+U$ rather than any other function of $T+U$, when the property which characterized $T+U$ has disappeared, that, namely, of being the sum of two terms of a particular form.

But this is not all; it is necessary to take account, not only of mechanical energy properly so called, but of the other forms of energy, heat, chemical energy, electric energy, etc. The principle of the conservation of energy should be written

$$T+U+Q=\text{constant}$$

where T would represent the sensible kinetic energy, U the potential energy of position, depending only on the position of the bodies, Q the internal molecular energy, under the thermal, chemic or electric form.

All would go well if these three terms were absolutely distinct, if T were proportional to the square of the velocities, U independent of these velocities and of the state of the bodies, Q independent of the velocities and of the positions of the bodies and dependent only on their internal state.

The expression for the energy could be resolved only in one single way into three terms of this form.

But this is not the case; consider electrified bodies; the electrostatic energy due to their mutual action will evidently depend upon their charge, that is to say, on their state; but it will equally depend upon their position. If these bodies are in motion, they will act upon one another electrodynamically and the electrodynamic energy will depend not only upon their state and their position, but upon their velocities.

We therefore no longer have any means of making the separation of the terms which should make part of T, of U and of Q, and of separating the three parts of energy.

If $(T+U+Q)$ is constant so is any function $\gamma(T+U+Q)$.

If $T+U+Q$ were of the particular form I have above considered, no ambiguity would result; among the functions $\gamma(T+U+Q)$ which remain constant, there would only be one of this particular form, and that I should convene to call energy.

But as I have said, this is not rigorously the case; among the functions which remain constant, there is none which can be put rigorously under this particular form; hence, how choose among them the one which should be called energy? We no longer have anything to guide us in our choice.

There only remains for us one enunciation of the principle of the conservation of energy. There is something which remains constant. Under this form it is in its turn out of the reach of experiment and reduces to a sort of tautology.[41]

But there are additional ambiguities. We talk of the sum of energies in the universe being constant, and this cannot even theoretically be supposed to be testable; on the other hand if we talk about any isolated system other than the universe, what do we mean by that? Is it not true that the only isolated system is the universe itself? If we choose to understand by the principle of conservation, that energy is indestructible, then we must always assume conversion into another kind of energy; but we can never be in a position to know that there are no further kinds of energy (as Poincaré pointed out above). All these and many more difficulties are cited in a very interesting paper by M. T. Keeton, 'Some Ambiguities in the Theory of the Conservation of Energy'.[42] Most of these difficulties occurred as a direct result of the establishment of the principle as a completely general one. The difficulties are valid, but so far, in the actual functioning of science, they have hardly exerted a disturbing influence, and even Poincaré's tautological definition of energy as 'that something which is being conserved' in nature, hardly caused trouble.

Conservation of energy: Cavendish

A well-nigh exact formulation of the conservation of energy appears among the posthumously published papers of Henry Cavendish; it is of special interest, since it came so close and did not succeed completely.

Among the 'Unpublished Papers of Cavendish' there is one titled 'Remarks on the Theory of Motion'.[43] The commentator on these papers,

Joseph Larmor, dates it at around 1760. A clear intellectual struggle is found when Cavendish attempts to distinguish between a momentum with direction, which he calls 'ordinary momentum' and that element of it which does not depend on direction and is called by Cavendish 'mechanical momentum'. He gives the following example:

From hence appears the nature of the dispute concerning the force of bodies in motion; for if you measure this force by the pressure multiplied into the time during which it acts the quantity of force which a moving body will overcome or the force requisite to put a body in motion or in other words the force of a body in motion will be as the velocity multiplied into the quantity of matter, but if you measure it by the pressure multiplied into the space through which it acts upon the body the force of a body in motion will be as the square of velocity multiplied into quantity of matter.

The 1st way of computing the force of bodies in motion is most convenient in most philosophical enquiries but the other is also very often of use, as the total effect which a body in motion will have in any mechanical purposes is as the quantity of matter multiplied into the square of its velocity; for in all mechanical purposes the force must be measured by the weight or resistance to be overcome multiplied into the height to which it is raised or the space through which it is moved, thus it requires an equal force to raise the weight of one pound 2 yards as it does to raise 2 pound 1 yard for the same force which is employed to raise 1 pound to 2 yards will by a proper machine raise 2 pounds 1 yard height. What I have here said will appear plainer by examples.

If a body moving with 1 degree of velocity is able to compress 1 spring to a certain distance the same body moving with 2 degrees of velocity will compress 4 springs to the same distance.

If in any machine a weight by descending communicates any degree of motion to the parts of the machine as in fig. 1st (p. 428) where the weight (a) is suspended by the string AB which is wound round the axis BC so that the weight cannot descend without putting in motion the fly dfg, then if the weight in descending any given height can make the fly revolve with any given velocity it will require 4 times that weight descending from the same height or the same weight descending from 4 times that height to make the fly revolve with 2^{ce} that velocity; when it is required to make the fly revolve with 2^{ce} the velocity by 4 times the weight descending from the same height this will not be exactly true, because as part of the weight is employed in accelerating itself there is then a greater weight to be accelerated than in the first case.

For like manner if one man can by applying his strength in the most advantageous [way] communicate a given velocity to any machine it will require 4 men to work during the same time to communicate 2^{ce} that motion to the engine, or it will require the force of one man for 4 times that time.

It must be observed here that as a man or any other cannot move with more than a certain velocity, and as the faster he moves the less force he is able to turn the engine with, therefore when it is required to move the engine with 2 degrees of velocity the velocity with which the man moves should be only $\frac{1}{2}$ as great in respect to that of the engine as when it is to be moved with only 1 degree of velocity.

The thickness of wall or timber which a cannon ball will force its way through is as the square of the velocity.

The work which an engine turn[ed] by a stream of water will do is as the quantity of water which strikes the wheel multiplied into the square of the velocity; for the pressure exerted upon the wheel is as the quantity of water × its velocity and the velocity with which the wheel may turn is as the velocity of the stream.

As this way of computing the forces of bodies in motion is very often of service, and as some theorems of considerable use in philosophy may be deduced from it, I would have some name by which to distinguish this way from the other; because it expresses the effect which a moving body will produce in mechanical purposes. I think it would not be amiss if it was called the mechanical force or mechanical momentum of bodies in motion.

When 2 perfectly elastick bodies of whatever shape strike each other in any manner whatever the sum of their mechanical momenta, not computed in one given direction but in any direction whatsoever, will be the same after the stroke as before.

When you say that the momenta of 2 bodies computed by the mass multiplied in the velocity are the same after the stroke as before you consider it only as made in a given direction, therefore if a body moves in a contrary direction you look upon its momentum as negative, if at right angles to it as nothing; but in this case whatever is the direction of a body's motion I still look upon it as positive and the same as if it was made in any other direction.

There is no need here of running into disquisitions concerning the nature of absolute and relative motion; for suppose any given point to be at rest and compute the motion of your system of bodies as they are in respect to that point, the truth of the proposition will be the same whether that point is really at rest or moving uniformly forwards in a right line.

Because in the collision of elastick bodies there is very often an encrease of momentum as usually computed some people have thought there might be a perpetual motion made from it, and indeed it at first seems very likely that it might be employed in mechanical performances to increase the force.

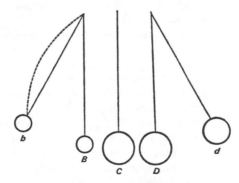

Thus suppose B, C, D to be any number of elastick balls encreasing in size from B to D let the ball B be drawn to any distance b and let fall again it will strike C and then D and the body D will move after the stroke with a much greater momentum than what B acquired by the fall; it should seem therefore that if when D was arrived at its greatest height d it should be stopt by a catch in such manner that its weight might be employed in moving the engine, that it would move it with more force than what was required to raise the ball B. But on the contrary it appears by the foregoing proposition that if the motion of all the other balls after the stroke could be in like manner applied to move the engine that their effect would be but rarely equal to what was required to raise the ball B.

It appears also from this proposition that whenever 2 nonelastick or imperfectly elastick bodies strike there is a loss of mechanical momentum; upon supposition that the bodies are intirely void of elasticity and strike each other directly the quantity which is lost may thus be calculated. Let the weight of the 2 bodies be A and B and their respective velocities $x+B$ and $x-A$ as before; after the stroke they will move both together with the velocity x and the sum of their mechanical moments will be $x^2 \times \overline{A+B}$, subtract that from their momenta before the stroke and there will remain

$$A^2B + B^2A = \frac{AB}{A+B} \times A + B^2.$$

If any point B instead of being considered as immoveable is in motion then the encrease of mechanical momentum produced in A by the attraction of B above what it would acquire by the attraction or repulsion of the other bodies added to that produced in B by the same attraction is equal to that produced in a body in falling by the same attraction through the space by which those 2 bodies approach one another.

Therefore if there is a system of bodies A, B, C, D, E & a attracting or repelling each other in the abovemention'd manner compute the mechanical momentum which it could produce in or take away from B in falling from B to A and also the momentum which it could produce in C in falling from C to A as also what it could produce by the same means in all the other bodies D, E, & a: in like manner compute the momentum which the attraction of B could produce in all the bodies which it attracts except A (as the momentum produced by their mutual attraction was before computed) compute also the momentum which C could produce in all the bodies it acts upon except A and B and do the same thing by all the other bodies, then the sum of these additional mechanical momenta added to the real momenta with which the bodies are moving will remain constantly the same and will not be altered by their actions upon one another.

Cor 2nd. Heat most likely is the vibrating of the particles of which bodies are composed backwards and forwards amongst themselves; therefore if bodies are composed of particles attracting or repelling one another in the manner above described their heat must remain constantly the same except as far as it is altered by receiving from or communicating heat to other bodies, and whenever 2 bodies of different heats are mixed together or otherwise placed so that one may receive heat from the other, one will receive as much mechanical momentum or in other words as great an encrease of heat multiplied into its quantity of matter as the other loses so that the sum of their mechanical momenta may remain unaltered. But there is plainly both an encrease and loss.

And in general we may conclude that whenever any system of bodies is in motion in such a manner that there can be no force lost by friction imperfect elasticity or the impinging of unelastick bodies, that then the sum of the mechanical momenta of the moving bodies added to the sum of the above-mentioned additional momenta will remain constantly the same.[44]

These long quotations speak for themselves. We see how a great scientist like Cavendish is struggling with concepts and contradictory experimental evidence to try to formulate an adequate mechanical conservation theory. We see that he could not have been acquainted with the work of d'Alembert—not unusual in England in those days; we see that he avoids carefully the expression '*vis viva*'; that in his struggle he is not helped by mathematical ideas which might make him choose a solution for the sake of its mathematical expediency. Moreover, his position is even more complicated by an almost clear statement of the motion theory of heat. In short, what we see is that the historical development had to be followed: first a clear mathematical formulation of the conservation principle in mechanics was needed, unhampered by not yet ready ideas from other fields of physics.

Conservation of energy: Newton

There is a widespread myth in history of science that Newton all but clearly enunciated the principle of conservation of energy. Though the Newtonian conception of 'force' is very complex, it clearly has very little to do with the concept of 'energy', and the origin of that myth remains to be examined.

In Thomson and Tait's *Lectures on Natural Philosophy* we find the following description:

229. What precedes is founded upon Newton's own comments on the third law, and the actions and reactions contemplated are simple forces. In the scholium appended, he makes the following remarkable statement, introducing another specification of actions and reactions subject to his third law, the full meaning of which seems to have escaped the notice of commentators:

Si aestimetur agentis actio ex ejus vi et velocitate conjunctim; et similiter resistentis reactio aestimetur conjunctim ex ejus partium singularum velocitatibus et viribus resistendi ab earum attritione, cohaesione, pondere, et acceleratione oriundis; erunt actio et reactio, in omni instrumentorum usu, sibi in vicem semper aequales.

In a previous discussion Newton has shown what is to be understood by the velocity of a force or resistance; i.e., that it is the velocity of the point of application of the force resolved in the direction of the force, in fact proportional to the virtual velocity. Bearing this in mind, we may read the above statement as follows:

If the action of an agent be measured by the product of its force into its velocity; and if, similarly, the reaction of the resistance be measured by the velocities of its several parts into their several forces, whether these arise from friction, cohesion, weight, or acceleration;—action and reaction, in all combinations of machines, will be equal and opposite.[45]

In this passage, they actually conclude that Newton discovered d'Alembert's principle:

230. Newton, in the passage just quoted, points out that forces of resistance against acceleration are to be reckoned as reactions equal and opposite to the actions by which the acceleration is produced. Thus, if we consider any one material point of a system, its reaction against acceleration must be equal and opposite to the resultant of the forces which that point experiences, whether by the actions of other parts of the system upon it, or by the influence of matter not belonging to the system. In other words, it must be in equilibrium with these forces. Hence Newton's view amounts to this, that all the forces of the system, with the reactions against acceleration of the material points composing it, form groups of equilibrating systems for these points considered individually. Hence, by the principle of superposition of forces in equilibrium, all the forces acting on points of the system form, with the reactions against acceleration, an equilibrating set of forces on the whole system. This is the celebrated principle first explicitly stated, and very usefully applied, by d'Alembert in 1742, and still known by his name. We have seen, however, that it is very distinctly implied in Newton's own interpretation of his third law of motion.[46]

But they claim more than that:

240. The foundation of the abstract theory of energy is laid by Newton in an admirably distinct and compact manner in the sentence of his scholium already quoted (§229), in which he points out its application to 'mechanics'. The *actio agentis*, as he defines it, which is evidently equivalent to the product of the effective component of the force, into the velocity of the point on which it acts, is simply, in modern English phraseology, the rate at which the agent works. The subject for measurement here is precisely the same as that for which Watt, a hundred years later, introduced the practical unit of a 'Horsepower,' or the rate at which an agent works when overcoming 33,000 times the weight of a pound through the space of a foot in a minute; that is, producing 550 foot-pounds of work per second. The unit, however, which is most generally convenient is that which Newton's definition implies, namely, the rate of doing work in which the unit of energy is produced in the unit of time.

241. Looking at Newton's words (§229) in this light, we see that they may be logically converted into the following form:

Work done on any system of bodies (in Newton's statement, the parts of any machine) has its equivalent in work done against friction, molecular forces, or gravity, if there be no acceleration; but if there be acceleration, part of the work is expended in overcoming the resistance to acceleration, and the additional kinetic energy developed is equivalent to the work so spent. This is evident from §180.

When part of the work is done against molecular forces, as in bending a spring; or against gravity, as in raising a weight; the recoil of the spring, and the fall of the weight, are capable at any future time, of reproducing the work originally expended (§207). But in Newton's day, and long afterwards, it was supposed that work was absolutely lost by friction, and, indeed, this statement is still to be found even in recent authoritative treatises. But we must defer the examination of this point till we consider in its modern form the principle of Conservation of Energy.[47]

The absurdity of this interpretation jumps to the eye. But for further emphasis let us quote a wonderfully naïve statement by Tait's biographer C. G. Knott:

I have heard Tait tell the story of the search after this interpretation. 'The Conservation of Energy', he said to Thomson one day, 'must be in Newton somewhere if we can only find it'. They set themselves to re-read carefully the *Principia* in the original Latin, and ere long discovered the treasure in the finishing sentences of the Scholium to Lex III.[48]

Back to Helmholtz

According to his own evidence, Helmholtz read in very early youth the works of d'Alembert, Euler, Lagrange and (naturally) Newton. Thus, before his professional training in medicine, the literature he read was mainly physics, philosophy and mathematics. As we have seen, the nineteenth century has inherited two basically different traditions in mechanics: Newtonian–vectorial mechanics, with its emphasis on forces, and the Leibniz–Euler–Lagrange formulation of analytical mechanics with its emphasis on the scalar quantities of the '*vis viva*' and the 'potential function'. The major concern of vectorial mechanics as formulated by the Newtonians was to measure the action of a force by its momentum—an approach that originated with Descartes. Descartes's momentum, however, undoubtedly is a scalar quantity and as such serves as a foundation to both traditions. The basic concepts in Newtonian mechanics were space, time, mass and force.[49] The drawbacks of this formulation were that for cases where constraints occurred the treatment became rather tedious; besides, the action-reaction law does not embrace all cases: it proves to be sufficient actually only for the dynamics of rigid bodies. Its great advantage was that forces, which are not derivable from a work function—i.e. which are not conservative but of a frictional nature, which cannot be dealt with by the mechanical conservation principle—can easily be treated by Newtonian mechanics. Needless to say, the apparent loss of *vis viva* in frictional processes troubled those physicists far less, for they could still solve the basic dynamical problem which they set themselves. On the other hand, the basic concepts in the Euler–Lagrange procedure were space, time, mass and energy. This procedure is only applicable to forces which are conservative, that is to say depend only on position and not on time or velocity. Here the conservation of mechanical energy holds—these are conservative systems. The great generality of the scalar treatment of mechanics becomes evident only if Hamilton's principle is introduced which covers all cases, even if the work function is a function of time. If action is defined as the integral of the difference between the kinetic energy and the work function, the principle says: 'the actual motion realized in Nature is that particular motion for which the action assumes the smallest value'. In all probability Helmholtz did not know of Hamilton's work at this stage (that is why I did not think it necessary to deal with it). But the reasons for not having taken this into account go deeper. The science of 'energetics' was a conceptually different structure from classical mechanics. This conceptual framework (of energetics) had at its foundation *two* basic principles: the principle of conservation of energy and the principle of least action. Its great advantage over classical mechanics was that the two principles taken together teach us more than the fundamental principles of the classical theory, and primarily they exclude motions not realized in nature which are compatible with the classical theory. But here logical development is not parallel to historical development of the science. In order to realize the importance of excluding processes which do not actually take place in nature, though admitted by classical mechanics, it was absolutely necessary to have realized the two fundamental principles of thermodynamics, i.e. the clarification of the 'entropy'—conserving character of complete reversible cycles.[50]

In 1847, Helmholtz, this being his first work in physics, was still struggling with the original Lagrangian formulation of mechanics and trying to reconcile it with the vectorial force-treatment. Although Helmholtz was a 'mechanical philosopher' and a great supporter of atomistic theories, he must have realized and appreciated the methodological advantages of the analytical approach: here constraints are dealt with in an elegant and easy way and there is no need to make any hypotheses concerning forces of constraint. On the other hand mathematically the two formulations of mechanics are equivalent and Helmholtz and his contemporaries must have asked the fundamental question, which I shall quote here, with the answer given again in the words of Lanczos:

Since motion by its very nature is a *directed* phenomenon, it seems puzzling that two scalar quantities should be sufficient to determine motion. The energy theorem which states that the sum of the kinetic and potential energies remains unchanged during the motion, yields only one equation while the motion of a single particle in space requires three equations; in the case of mechanical systems composed of two or more particles the discrepancies become even greater. And yet it is a fact that these two fundamental scalars contain the complete dynamics of even the most complicated material systems, provided they are used as the basis of a *principle* rather than of an equation.[51]

It need hardly be emphasized, that the principle referred to is that of least action—the great unifying principle of nature, metaphysically postulated by Maupertuis, mathematically formulated by Euler and Lagrange and finally generalized by Hamilton. In 1847 Helmholtz could not have given this answer.

Before taking up what Helmholtz actually did, we have to pursue two further avenues of approach: the tradition in the theory of heat, and the physiological background.

NOTES

1. In the *Physics* (Ross's edition), bk. iii, 201.a, 10.
2. ibid.
3. R. J. Blackwell, R. J. Spath and W. E. Thirkell, *Commentary on Aristotle's Physics by St Thomas Aquinas* (Yale, 1963), p. 137.
4. Stephen Mackenna (Ed.), *Plotinus' Psychic and Physical Treatises* (vol. ii, London, 1921), p. 196.
5. S. Sambursky, *The Physical World of Late Antiquity* (London, 1962), p. 106.
6. J. A. Weisheipl, *The Development of Physical Theory in the Middle Ages* (New York, 1959), p. 48.
7. It does recur in Leibniz and the philosophers who were influenced by him, but in a modified way.
8. G. W. Leibniz, 'On Nature itself, or on the Inherent Force and Actions of Created Things'. The original Latin appeared in *Acta Eruditorum* (1698).
9. L. Loemker (ed.), *Leibniz's Philosophical Papers and Letters* (2 vols., Chicago, 1955), p. 808.
10. We shall return to this elsewhere, but at this point the author would like to advance the opinion that the claims about Leibniz's complexity and incomprehensibility seem to be a result of sheer hindsight: Leibniz's aims and techniques are crystal clear; his final solution of the unity of mind and matter, incorporating his dynamics, is necessarily vague and he knew that. It is not the monads that have to be understood clearly, but the problem for the solution of which they were introduced—the monad was not meant to be clear and translatable into physical terms; the situation is somewhat similar to the Anaximandrian 'nous' or the Kantian 'force' or the Naturphilosophs' 'nature' or the nineteenth-century physiologists' 'vital principle' or, perhaps, to the twentieth-century 'elementary particle'. The very importance of these concepts is that these are the terms in which the philosopher-scientists think before they succeed in completing a philosophical or physical system, and thus the terms in which they could be made clear do not yet exist. Thus every attempt to make these concepts clear is an attempt to translate them into a language which presupposes the acceptance of a conceptual framework, the overthrow of which was the primary reason for introducing a new concept. To prove that this theory applies to the concept of energy too, is one of the main purposes of this thesis.
11. The quotation is taken from the *O.E.D.*, under 'energy'.
12. Thomas Young, *Lectures on Natural Philosophy* (2 vols., London, 1807).
13. ibid.
14. E. J. Dijksterhuis, *The Mechanization of the World Picture* (Oxford, 1970), p. 411.
15. J. Herivel, *The Background to Newton's Principia* (Oxford, 1965), p. 151, n. 4.
16. W. L. Scott, 'The Significance of "Hard Bodies" in the History of Scientific Thought', *Isis 50* (1959), pp. 199–210.

17. J. le R. d'Alembert, 'Energie', entry in the *Grande Encyclopédie*. On this see T. Hankins, *Jean d'Alembert* (Clarendon Press, Oxford, 1970), ch. 9.

18. *The New Physical Hypothesis* (1691), which appeared in two parts: (1) The Theory of Abstract Motion and (2) The Theory of Concrete Motion.

19. There is the famous quotation by H. A. Kramers which I have already used as the motto for this book, but it is so poignant that it is worth repeating:

In the world of human thought generally and in physical science particularly, the most fruitful concepts are those to which it is impossible to attach a well-defined meaning.

20. Jean Nic. Seb. Allamond, Ed., *Œuvres Philosophiques de Mr G. G.'s Gravesande* Amsterdam, 1774), p. 305, 'Remarques Touchant le Mouvement Perpetuel'.

21. On this I refer to Thomas L. Hankins (1965)

22. De Miran, 'Dissertation sur l'estimation et la mesure des forces motrices des corps' (1728).

23. The original was published in the *Historie de l'Académie Royale des Sciences* (Paris t.4, 1775), pp. 61–6. The translation here is taken from H. Dricks, *Perpetual Motion in the 17th and 18th Centuries* (London, 1861), p. 190.

24. J. P. Lagrange, *Mécanique Analytique*, I, p. 19.

25. E. Hiebert, *Historical Roots of the Principle of Conservation of Energy* (Madison, 1962), p. 95.

26. C. Truesdell, 'A Program toward Rediscovering the Rational Mechanics of the Age of Reason', *Arch. Hist. Exact Sci.* **1** (1955), 11.

27. L. Euler, 'Mechanica sive Motus Scientia Analytica Exposita', ed. P. Staeckel, in *Opera Omnia Series Secunda*, t. 1.

28. ibid.

29. More on Euler see in ch. I, note 8.

30. J. Le R. d'Alembert, *Traité de Dynamique*, Ed. Gauthiers-Villars (1921).

31. R. Dugas, *History of Mechanics* (Geneva, 1955), p. 247.

32. Lagrange, *Mécanique Analytique*, p. 224.

33. D'Alembert, *Traité de Dynamique*, p. xxvi.

34. ibid., p. xxviii.

35. Hankins, *Jean d'Alembert*.

36. C. Lanczos, *The Variational Principle in Mechanics* (Toronto, 1954), p. 90.

37. For a clear treatment, see Lanczos, ch. viii, p. 193.

38. Dugas, *History of Mechanics*, p. 343.

39. ibid.

40. E. Mach, *History and Root of the Principle of Conservation of Energy*, trans. P. E. B. Jourdain (Open Court Publishing Co, 1911).

41. H. Poincaré, *Science and Hypothesis*, ch. viii, 'Energy and Thermodynamics' (Dover, 1952).

42. M. T. Keeton, 'Some Ambiguities in the Theory of the Conservation of Energy', *Phil. Sci.* **8** (1940), 304.

43. Sir E. Thorpe, Ed., *The Unpublished Scientific Papers of Henry Cavendish* (Cambridge, 1921).

44. ibid.

45. W. Thomson and P. G. Tait, *Lectures on Natural Philosophy* (Cambridge, 1879).

46. ibid.

47. ibid.

48. C. G. Knott, *The Scientific Work of P. G. Tait* (Cambridge, 1911), p. 191.

49. Thompson and Tait, *Lectures*.

50. Carnot certainly did not have a clear idea about the conservation of energy

principle. As to what he meant by 'calorique' as against 'chaleur', an interesting discussion was conducted in the *American Journal of Physics*, some years ago: (1) V. K. La Mer, 'Some current misinterpretations of N. L. Sadi Carnot's Memoir and Cycle', *Amer. J. Phys.* **21** (1953), 20; (2) Thomas S. Kuhn, 'Carnot's Version of "Carnot's Cycle",' *Amer. J. Phys.* **22** (1954), 91; (3) V. K. La Mer, 'Some Current Misinterpretations of N. L. Sadi Carnot's Memoir and Cycle II', *Amer. J. Phys.* **22** (1954), 95; and also the following: F. O. Koenig, 'On the History of Science and of the Second Law of Thermodynamics' in H. M. Evans (ed.), *Men and Moments in the History of Science* (University of Washington Press, 1959); S. Lilley, 'Attitudes to the nature of heat about the beginning of the nineteenth century', *Arch. Intern. d'Histoire des Sciences* **27** (1948), 630. The connection between Carnot's various concepts and that of energy (that is after the proof of the principle of its conservation) is also treated later on.

51. Lanczos, *Variational Principle*, p. xviii.

III

HEAT AND ENERGY

The two laws of thermodynamics

For the physicist the concept of energy is inseparable from the principle of its conservation. It is significant that all those historical studies where energy concepts are attributed to various natural philosophers in the seventeenth and eighteenth centuries are not written from the physicist's point of view. On the other hand, studies in the history of the concept written by physicists make it very clear that the concept of energy in its modern sense was born after the establishment of the principle of conservation. Let me emphasize this once more: by 'energy in its modern sense' I mean the classical, pre-relativity point of view, on the understanding that $E=h\nu$ created a new concept of energy, and so did the Einsteinian formula of mass–energy equivalence. The concept of energy in the general theory of relativity is again a conceptually new development, as it has been amply demonstrated in recent articles by Professor Møller.[1] The establishment of the principle of conservation of energy led to the science of 'energetics', rich in metaphysical and even religious overtones. Thermodynamics began on a much more modest scale, and today constitutes one of the few cornerstones of physics not shattered by twentieth-century developments. I shall briefly review the two fundamental principles of thermodynamics in their modern form, to make the consecutive historical narrative more comprehensive. But already at this stage a clarification is necessary. In thermodynamics a clear distinction has to be drawn between reversible and irreversible processes, a distinction which is neither predicted nor permitted in any other part of physics like mechanics or electrodynamics. Many textbooks and scientific articles have been written in the last decade, and more are appearing daily, precisely with the aim of describing this distinction, as it constitutes one of the open problems of science. We shall naturally confine the discussion to reversible processes. In other words, thermodynamics as presented here does not apply to a great many of the processes of life; needless to say this qualification described in such clear terms is of rather recent date. A. Katchalsky and Peter F. Curran have recently published a book called *Nonequilibrium Thermodynamics in Biophysics*; in the introduction they say:

Classical thermodynamics is, however, limited in its scope. It is essentially a theory describing systems in equilibrium or undergoing reversible processes and is particularly applicable to closed systems. . . . In the case of such irreversible processes, the laws of classical thermodynamics provide a set of inequalities describing only the direction of change.[2]

The latest work in this field has been to replace the above-mentioned inequalities of classical thermodynamics with equalities which now make a quantitative description of irreversible processes possible.

Thermodynamics deals essentially with macroscopic phenomena and its task is to describe the world in terms which are derived from these phenomena. As the 'world' as a whole cannot experimentally be investigated, certain parts of it are artificially isolated under controlled experimental conditions; such isolated parts are called 'thermodynamic systems' or just 'systems', and we are dealing with these. If all exchange of matter or thermal energy between the system and its surroundings is prevented, the system is designated as an 'adiabatic system'. If such an adiabatic system is subjected to the operation of external forces X_i which cause the external parameters of the system a_i to change by an amount da_i, a quantity of work dW is performed according to the equation

$$dW = \sum_{i=1}^{n} X_i da_i.$$

If the force concerned is isobaric pressure P, and as a result of its application the volume changes by dV the work performed will be $dW = PdV$. It is experimentally proved (originally by the various experiments on equivalence between heat and mechanical work, performed with the greatest exactitude by Joule) that the work W_{12} accompanying a transition of an adiabatic system from one state to a second is independent of the physical path. In the above case, where the work done is compression, the performance will be

$$W_{12} = \int_{1}^{2} PdV.$$

When taking the adiabatic system from state 1 to state 2, the work is independent of the intermediate values assumed by P and is fully determined by the initial and final states of the system.

The considerations for an adiabatic system can easily be generalized so as to include a closed system separated by a diathermal wall from a heat reservoir. If we now enclose the heat reservoir and our system in an adiabatic wall we can assign to the whole system (closed system plus reservoir) an internal energy U, which is the sum of the internal energy of the closed system A, i.e., U_A and of the internal energy of the reservoir U_B. As we have seen the total work done is independent of the path, that is dW is a total differential; in other words there exists a potential function of state whose differential decrease represents the element of work. This is what we called internal energy and

$$- dU = dW$$
$$\text{or} \quad dU_A = - dU_B - dW.$$

Since the whole system is adiabatic, the energy decrease in B (the reservoir) must be due to transport of energy from B to A. This transported energy is heat and shall be designated by Q. Thus for our system $dQ = dU_B$ and we get

$$dU_A = dQ - dW.$$

This is the mathematical statement of the first law of thermodynamics, as originally formulated by Helmholtz. (The physical definitions and formulation are taken almost verbatim from the above-quoted excellent new textbook on irreversible thermodynamics by Katchalsky and Curran.)

The law as formulated by Helmholtz is still valid; it applies to relativistic processes as well as to all processes described by quantum mechanics. As with all conservation laws, it also applies to impossible as well as to possible processes in nature. This was the growing realization in the 1850s which made the already independently discovered Carnot principle the other fundamental principle of thermodynamics. The first law implies the impossibility of a *perpetuum mobile*, and it implies the mutual convertibility of all forms of energy; but does not exclude processes which are impossible in nature, and which are generally called perpetual motion machines of the second kind. In other words it does not take into account that natural and spontaneous processes exhibit a directional property. While idealized processes are fully reversible, natural processes always proceed towards an equilibrium position and on reaching it the process stops. In other words, spontaneous processes dissipate their motive power. This was the phenomenon named by William Thomson the 'dissipation of energy', and the phenomenon that occupied the attention of Clausius for many years. This directional property of natural processes was taken into account in what came to be known as the second law of thermodynamics in one or the other of its numerous formulations.

Lord Kelvin's formulation follows:

It is impossible by means of an inanimate material agency to derive mechanical effect from any portion of matter by cooling it below the temperature of the coldest of the surrounding objects.[3]

In mathematical terms

$$\oint (\mathrm{d}W)_T \leq 0,$$

where the subscript T means that throughout the cycle the temperature is a constant. For reversible processes the '$\leq$' would simply be replaced by '$=$'. The usual alternative formulation of the second law is that of Clausius:

It is impossible to construct a device that, operating in a cycle, will produce no effect other than the transfer of heat from a colder to a hotter body.[4]

In the last century, before it was realized that thermodynamics is a more general theory than mechanics, there was endless activity among the physicists to try to reduce the two fundamental laws of thermodynamics to the laws of mechanics. It has been shown often (in particular by Poincaré) that a hypothesis of molecular forces is enough to deduce the law of conservation of energy from the laws of mechanics; in a way it was also done by Helmholtz in his 1847 paper. There were greater difficulties with the second law. Clausius tried it without success, and again Helmholtz succeeded to a much greater extent in his final work *On the Principle of Least Action*. Naturally this theory was not complete either, for it does not cover irreversible processes at all. Slowly this desperate attempt to reduce thermodynamics, and with it the whole physical world, to phenomena of mechanics, gave place to a new autonomous science, thermodynamics, which stands strong today on the foundations outlined above.

A distorted historical picture and a suggested remedy

The physical order of things very often prejudices our historical view. Because of the above-described physical state of affairs, there is an oversimplified story of the historical development of thermodynamics and the theory of energy; the argument is somewhat as follows. Thermodynamics is based on two fundamental principles; the first of these is the law of conservation of energy which is both historically and logically prior to the second law (i.e. Carnot's principle in one of its earlier or later formulations). The second law was discovered by Sadi Carnot in the 1820s and thus the first law must have been discovered earlier. Indeed, if we look carefully, we shall find that there was a well-established law of conservation of mechanical energy, first formulated by Huygens and Leibniz. Some historians and scientists even agree with Tait and Thomson that Newton already had an almost clear formulation of the principle. As in the late eighteenth century difficulties with the phlogiston theory led to its gradual replacement by the theory of the caloric, this led to new obstacles in the way of a mechanical theory of heat, which, as everybody knows, was the last necessary step before a most general principle of conservation of energy could be established. Now, both Rumford and Davy were champions of the theory that heat was motion, but unlike their predecessors Bacon, Boyle and Locke, these two proved it experimentally and conclusively. After that all was clear: heat is motion; motion is kinetic energy; the sum of kinetic and potential energies is a constant, and thus the first law of thermodynamics has been established.

This picture has all the lucidity and simplicity that one can require, the only disturbing element about it being that it is historically false. Historians of science have been aware for some time now that not all is well with the above argument, but the problem has not yet been explored in depth. It was this vague feeling of something's being basically wrong that led to the numerous discussions on the exact nature of Carnot's work and on the concept of his 'calorique', mainly between Professors T. S. Kuhn[5] and Victor K. La Mer.[6] The oversimplified picture has also been vigorously attacked by S. Lilley in various publications, especially in his 'Attitudes to the Nature of Heat about the Beginning of the Nineteenth Century'.[7] This is a complicated problem, and will be dealt with only so far as it is inseparably connected with our own problem.

At this stage several problems rate attention and have to be settled before the whole story can be told. Before any historico-physical studies are initiated, it must be made clear that scientific concepts do not emerge in a logical, clear-cut-for-the-need form; that whatever the final formulation of a physical law, the concept that emerges of this formulation is not identical with the concepts in which the discoverer of that law was working while his discovery was being crystallized; any such attempt of identification is pure hindsight. The author attempted this sort of clarification in the introductory chapter, and adapted it to the concept of energy in this whole study. A full-length study is required for many other related concepts. One of the first of these is the concept of 'caloric' or Carnot's 'calorique'; this would imply a sympathetic and scientific study of the caloric theory, on the presupposition that scientists of all periods were not less clever or ingenious than we are. It is thus necessary to examine exactly which of the then known physical phenomena could be explained by the caloric theory and which could not, as against which well-known phenomena required a theory that heat was motion. It is by no means less important to review separately those phenomena which admit both explanations and enquire for the criteria which could have been advanced for a clear-cut decision. One of the major clarifications that would then immediately emerge is that there is no immediate connection between the motion theory of heat and any ideas of conservation; this leads us into an exhaustive study of how and which *a priori* conservation ideas created physical science, and what the relation is between the various entities which used to be considered conserved in various times, and those entities which are considered as conserved in our physical science. These and many more. All this is mentioned only to give a background against which, or as part of which, I would like this study to be seen. Much of the above-mentioned topics have been dealt with lately, but not systematically enough. There are the illuminating studies of Professor Sanborn Brown on the caloric theory;[8] one can wish only that they were double the length. There is the long study by Professor Koenig on the history of the second law;[9] he has been working on a comprehensive history of thermodynamics for many years. Professor E. Hiebert wrote 'The History and Roots of the Principle of the Conservation of Energy'[10] with the same purpose in mind. And what is to my mind even more important, one should re-read and re-evaluate the conceptual analyses of Duhem and Poincaré. This leads us into another type of work which began with scholars like Meyerson, Cassirer, Brunschvicg and Koyré in our century, and Whewell, Comte and Duhem in the last: what I have in mind is the interaction between science and philosophy—every great scientist and every great philosopher has predecessors as well as pupils both among scientists and among philosophers. This interaction is the most fruitful in the development of human thought. We cannot enlarge upon this theme here, but at least for the concept of energy, the author will make this attempt in the relevant sections.

We saw in the previous chapter that there was a mechanical law of conservation of the sum of *vis viva* and potential 'energy'. We saw the power and the limitations of this law; the extent of the gulf between the theoretical science of mechanics and the experimental science of physics was also stressed. To assume that there must have been some attempt to unify all these fields is again hindsight—one of the chief results of the establishment of the principle of conservation of energy. Only after its formulation and after the foundation of the all-embracing science of thermodynamics, did the sense of unity of all natural phenomena—that is the reintroduction of mechanics into physics—become a part of the working physicist's mental equipment. The emphasis here is on the working physicist, for the philosopher-physicist has long been committed to this, although he never paid sufficient attention to those experimental data which clashed with his special model of the unity of nature. The mechanician-mathematician of the late eighteenth century, taught and indoctrinated by the spirit which became embodied in his École Polytéchnique, built and rebuilt corpuscular or elastic-fluid models for most of the phenomena he dealt with, but certainly did not deal with conservation principles, except in the limited (in our sense) field of mechanics. Again, this is a problem which cannot be dealt with simply by calling names: the French mathematical school were 'positivists' and thus did not really believe in the reality of physical models, etc. It is still an open question how Cauchy could have constructed one corpuscular theory of elastic matter and one continuum theory without coming up with unanswerable riddles and on what methodological foundation does the joint work of Lavoisier and Laplace rest?

We shall now take up briefly the problems of the caloric theory as it stood in the days of Lavoisier, examine the physical and philosophical stand of Rumford and Davy, and then via Carnot reach the point where we can say 'this is the heritage received from the physical theory of heat'.

The old motion theory of heat

The old theory on the nature of heat was dynamical, and its chief representatives were Bacon, Boyle, Hooke and Locke. On a completely different level, influential mainly on the continent—though not of immediate import—was Leibniz's conception that heat was motion. Bacon had expressed himself very clearly on the subject. However, until the end of the eighteenth century, when Rumford and Davy suddenly began experimenting on the basis of these theories, not many physicists took them seriously. That the old theory had great influence on physiological theory we shall see later; it was spread mainly through the works of Albert Haller and Boerhaave.

In the twentieth aphorism of the Second Book of the *Novum Organum* we find the following statements:

When I say of motion that it is the genus of which heat is a species, I would be understood to mean, not that heat generates motion, or that motion generates heat (though both are true in certain cases) but that heat itself, its essence and quiddity, is motion, and nothing else. . . . Heat is an expansive motion, whereby a body strives to dilate and stretch itself to a larger sphere or dimension than it had previously occupied. . . . Heat is a motion of expansion, not uniformly of the whole body together, but in the smaller parts of it; and at the same time checked, repelled, and beaten back, so that the body acquires a motion alternative, perpetually quivering, striving and struggling, and irritated by repercussions, whence springs the fury of fire and heat. . . . And this specific difference is common also to the nature of cold; for in cold contractive motion is checked by a resisting tendency to expand just as in heat the expansive action is checked by a resisting tendency to contract. Thus whether the particles of a body work inward or outward, the mode of action is the same.[11]

The long quotation is justified, since generally in textbooks on the history of science, the name of Bacon is associated with the theory that heat is motion, but he is seldom quoted verbatim. However, this passage is particularly interesting, as it shows to what extent this theory was mere speculation, echoing the Greek atomists without having much physical content.[12]

Hooke was another important representative of the motion theory of heat. His approach was certainly more physical than that of Bacon, and indeed he was one of the few whom Rumford quoted as an authority on his ideas. (I have not undertaken a study of Hooke, and though in view of the approach developed here, I would not expect to find in his works a clear formulation of the principle of conservation of energy, there were some who did.) Louise Dahl Patterson published an article in *Isis* on 'Hooke and the Conservation of Energy'.[13] She certainly established a good case to prove that Hooke did have a conservation principle limited to mechanics; on the other hand it seems to me unjustified to claim that these theories (i.e. conservation of *vis viva* and that heat is motion) were connected for him except in the vaguest sense.

Tyndall quotes a passage from Locke, in his *Heat as a Mode of Motion*:

Heat is a very brisk agitation of the insensible parts of the object, which produce in us that sensation from whence we denominate the object hot; so what in our sensation is *heat*, in the object is nothing but *motion*.[14]

But the working physicists and chemists were not satisfied. They had to have a theory which would apply to expansion on heat, to changes of state and also to the rapidly accumulating information on chemical compounds and reactions; moreover it had to be a theory which accommodated the accepted corpuscular philosophy. One suggested theory was Stahl's Phlogiston Theory. This stage I shall skip here—its theory has been amply documented and in any case it is not intimately connected with the development of the concept of energy. On the other hand we have to take very seriously the caloric theory, which having been proposed by Lavoisier, together with the theory of latent heat by Black, had such enormous explanatory power that its very success prevented the unification of atomic theory in physics and the new chemical theory, a unification which turned out to be a conceptual *sine qua non* for the connection of mechanical 'energy' and phenomena of heat.

The caloric theory of heat

For most of the eighteenth century the material—and the motion—theories of heat enjoyed equal popularity. One can say that both retained the loyalty of their adherents on the basis of philosophical or traditional grounds, since neither of them was really a quantitative physical theory. This situation reflects the stagnation in the development of chemistry, while the best minds of the century were occupied either in carrying the achievements of Newtonian mechanics into the yet unexplored field of fluid mechanics and inventing better and better perturbation theories for the needs of astronomy, or carrying corpuscular philosophy, as formulated in the *Opticks*, into various new fields like elasticity and theory of light.

The two groups of scientists are not mutually exclusive; at their intersection we shall find most of the great French mathematicians. Activity in astronomy was nearing its zenith, and that in contemporary molecular physics was approaching an impasse. As an interesting twist of history, the resumption of physical inventiveness came first from the new science of chemistry, with the accompanying new material theory of heat brought to perfection, and only when using the results of Black, Lavoisier and other eminent chemists could the new science of thermodynamics be formulated. Only after the work of Clausius, when a true mechanical theory of heat had again been established, did the atomic theory of Dalton and his followers re-enter the mainstream of physics.

In the hands of Black and Lavoisier the material theory of heat developed into a quantitative science, and a serious attempt has been made to explain all phenomena of heat in terms of an elastic fluid, or as he also called it an 'igneous fluid'. And one can follow his mode of thought on these lines, even when he emphasizes that his theory of heat is only a plausible hypothesis; in any case this methodological statement is encountered in such joint enterprises as his essay 'Memoire sur La Chaleur', which he wrote together with Laplace in 1780.[15] Laplace was an ardent molecular physicist, and it is very probable that he in reality believed in the motion theory of heat. (The evidence as to Laplace is very scarce, and not enough study has been done on his work to decide on this question with certainty.) They write:

Les physiciens sont partagés sur la nature de la chaleur. Plusieurs d'entre eux la regardent comme un fluide répandu dans toute la nature, et dont les corps sont plus ou moins pénétrés, à raison de leur temperature et de leur disposition particulière à le retênîr. . . . D'autres physiciens pensent que la chaleur n'est que le resultat des mouvements insensibles des molecules de la matière. . . . C'est ce mouvement intestin qui, suivant les physiciens dont nous parlons, constitue la chaleur.[16]

After that, it is clearly stated that on this theory the motion is equal to the product of mass and the square of the velocity, that is with the *vis viva* for which the law of conservation of *vis viva* holds! And this argument does not seem important enough for the material theory to be discarded! It is just one of the numerous arguments for and against both hypotheses. And yet there is a conservation law here which might be considered the first one which is common to both hypotheses:

Quoi qu'il en soit, comme on ne peut former que ces deux hypothèses sur la nature de la chaleur, on doit admettre les principes qui leur sont communs; or, suivant l'un et l'autre, *la quantité de chaleur libre reste toujours la même dans le simple mélange des corps.*[17]

It is important to remember that Lavoisier was motivated mainly by his anti-phlogistic theory, and that the phlogistic theory was not a material theory of heat only slightly different from the 'caloric'; indeed Fourcroy, according to Lilley'[18] in his phlogistic days inclined towards the mechanical theory of heat and only during his very slow and gradual acceptance of anti-phlogistic chemistry did he go over to the material theory of heat.

Lavoisier used the word 'caloric' which was later adopted as the official term for the matter of heat in the chemical usage. Robert Hare, an ardent supporter of this theory, in the introduction to his *Compendium of the Course of Chemical Instruction*, notes the following:

It would follow from using the word heat in the sense both of cause and effect, that there is more heat in a cold body than in a hot one, which in language is a contradiction. On this account it was considered proper by the chemists of the Lavoisierian school, to use a new word, *caloric*, to designate the material cause of calorific repulsion.[19]

The 'caloric' was a fluid the particles of which were self-repulsive. This force counter-balanced their gravitational attraction and in fact prevented the collapse of all bodies into a solid homogeneous mass under the influence of this gravitation. This model supplied an obvious solution to problems like the expansion on heating and contraction on cooling. According to this theory each atom of the body is surrounded by a more or less dense caloric cloud, the density of which falls off as

$$\frac{1}{r^n} \text{ (where } n \text{ is larger than 2),}$$

that is faster than gravitational attraction, and thus it is a short-range force relative to the other one. Already in the 1730s Boerhaave and Fahrenheit mixed liquids at different temperatures and tried to evaluate the final temperatures: they took equal masses (or volumes) of water and found the final temperature as

$$T = \frac{T_1 + T_2}{2}.$$

That this was a special case of a general formula was slowly realized, and around 1750 the equation

$$T = \frac{m_1 T_1 + m_2 T_2}{m_1 + m_2}$$

became known. Professor Tisza[20] points out that this equation can be 'considered as the consequence of a conservation law' but I doubt that it is true that 'it has been derived on such a basis'. The question naturally fits into the framework of which model of heat are we looking at in the problem. The caloric theory posed no special problem with a conservation law, since the conservation of matter had been known for a long time, and Lavoisier's experiments and his insistence on the importance of mass in chemical calculations made the conservation of caloric easily acceptable.

In the work of Black several important aspects of the problem became clarified.[21] He introduced clearly what has become known as 'specific thermal capacity', the problem of change of phase by the new concept of 'latent heat' (as against the free heat) and a clear differentiation between the intensity of heat (from now on called 'temperature' in the modern sense) and the quantity of heat which became Lavoisier's 'caloric'. All this on the material theory of heat. Incidentally, Black who claims to have examined very carefully all the contemporary theories of heat, decided to reject the old motion theory of heat, the Boerhaave version of which supposed that 'the motion in which heat consists is not a tremor, or vibration of the particles of the hot body itself, but of the particles of a subtile, highly elastic, and penetrating fluid matter, which is contained in the pores of hot bodies or interposed among their particles'; he chose finally Cleghorn's theory which was announced later in the latter's *Disputatio Physica Inauguratis, Theoriam Ignis Competentur,*[22] Edinburgh, 1773. It is not clear to me who was the first to advocate it.

The new differentiation between temperature and quantity of heat by Black (on the basis of the material theory) is best illustrated by a diagram introduced in Sanborn Brown's article on caloric theory. I quote:

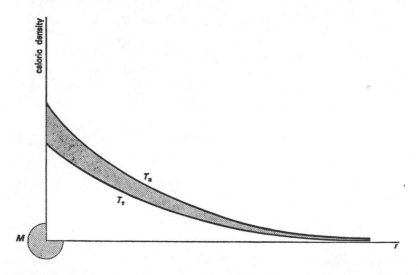

The two curves represent the caloric density at two different temperatures. The intensity of heat is represented by the temperature and therefore by the actual density of caloric at the surface of the atom. All atoms did not have identical caloric atmospheres, and although they all had logarithmic dependence of caloric density on distance, the rate at which the atmospheric density falls off varies from substance to substance. The quantity of heat required to change a body from temperature T_1 to temperature T_2 is represented by the shaded difference between the two curves.[23]

Two further discoveries were in full agreement with the caloric theory. In 1780 Laplace and Lavoisier in their joint mémoire showed that specific thermal capacities are not constants but functions of the temperature. Though this was not meant as a proof of the caloric theory, it agrees with it well, as on the caloric theory of specific heat, the density of the cloud of caloric around the atom is a function of the temperature. Dulong and Petit, in their joint article,[24] also predicted that the expansion on heating of a body should not be a uniform function of temperature, but rather that this should rise with the temperature. This was again in full agreement with the caloric theory. A year later they showed (and again I quote Professor Brown):

that the product of the specific heat of each solid element by the weight of its atom gave a constant value. This led to the conclusion that the amount of caloric surrounding an atom was related to its atomic weight in a manner which was expected from the attraction between matter and caloric.[25]

Latent heat was explained simply as a chemical combination between ordinary matter and the matter of heat.

We can sum up the conceptual structure of the physical theory of heat on the caloric model (following Professor Tisza) in two formulas. Defining heat quantity as the product of mass and temperature $\delta Q = m\delta t$ we obtain the conservation law

$$\delta Q_1 + \delta Q_2 = 0,$$

expressing the well-known result in elementary calorimetry that the heat gained by one body equals the heat lost by the other, and is actually equivalent to the above-mentioned formula

$$T = \frac{m_1 T_1 + m_2 T_2}{m_1 + m_2}.$$

The other equation will express the heat content of the body, with $c(T)$ being the thermal capacity:

$$Q(T) = Q(T_0) + \int_{T_1}^{T_2} c(T) dT$$

and

$$c(T) = c(T_0) + \sum_i L_i (T - T_i)$$

where T_0 is any arbitrary standard temperature, and L is the latent heat.

It is thus evident that the caloric theory has an enormous explanatory power covering most of the known phenomena of the time. It did not explain 'excitement' of heat on friction (as Rumford prefers to call it as against the caloric 'freeing of heat'); it runs into difficulties with radiation theories too; but we have to remember that these phenomena gained much in importance only fifty years later when the experiments with transformation of heat directly or indirectly into mechanical effects or the converse became popular. In the meantime it had won the day as far as the chemists were concerned.

Count Rumford

As mentioned above, Rumford and Davy are credited with the experimental proof of the theory that heat is motion. And yet it is fascinating to observe how little methodological or conceptual affinity there is between these two men. As we shall soon see Sir Humphry Davy was philosophically-minded and thus very interesting as such, while his scientific contribution to the motion theory of heat was negligible. Moreover, he was not even committed very much to this theory, and to the end of his life his doubts as to the real nature of heat only deepened. On the other hand Count Rumford's whole scientific career was devoted to this one topic, and untroubled by methodological and philosophical doubts he worked all his life consistently towards his goal. Were it not for my slavish dependence on historical order, I would have liked to deal with Rumford after having considered the work of Davy.

His *Mémoire sur la Chaleur*[26] begins with a 'Notice Historique': what he wrote there is very characteristic of the famous Count. Though admittedly in the eighteenth century, and the first half of the nineteenth too, scientists were not very conscientious about quoting their predecessors, yet, even in those times a 'Notice Historique' at the beginning of a scientific treatise would raise the expectation that a brief history of the subject is to follow. Not in this case however—we are only to be given a brief history of Count Rumford's own experiments and the list of his own discoveries. This being so, and in view of the fact that he never quotes other authorities, it is noteworthy that here the name of Boerhaave does come in very strongly:

Il y a très-long-temps que *la chaleur* fait l'objet favori de mes recherches. Ce fût en lisant l'excellent traité de Boerhaave sur le feu, a l'age de 17 ans, que je me suis attaché à ce sujet.[27]

This reference is even more significant because it was not accidental that Rumford never quoted other authorities. At the end of his lecture to the Royal Society on 'An Experimental Inquiry concerning the Source of the Heat which is Excited by Friction', held in 1798, he explains:

Before I finish this Essay, I would beg leave to observe, that although, in treating the subject I have endeavoured to investigate, I have made no mention of the names of those who have gone over the same ground before me, nor of the success of their labours; this omission has not been owing to any want of respect for my predecessors, but was merely to avoid prolixity, and to be more at liberty to pursue, without interruption, the natural train of my own ideas.[28]

Thus it is all the more important that Boerhaave was mentioned. Boerhaave's theory was that a heated body is like a struck bell, that is, it vibrates rapidly; this image stuck in young Benjamin Thompson's mind and became a leading principle of his researches on heat. There is a thorough analysis of Boerhaave's Doctrine of Fire in I. Bernard Cohen's *Franklin and Newton*.[29]

Thus, from the very early age of seventeen, Benjamin Thompson's interest in science was focused on the problem of heat, and with a strong prejudice in favour of a theory that heat actually resided in the vibratory motion of the material particles. By this I do not mean that he was committed to any sort of deeply influential metaphysical belief. He certainly did not have that kind of scientific temperament, and we do not find in his entire writings on scientific, social or military matters any remark which could thus be interpreted. What I mean is simply that as every scientist approaches a problem with some theory in mind (I hope that by now the myth of the scientist who goes into his laboratory with a blank mind to do experiments for eight hours, has been discarded by everybody), the theory that Thompson (the later Count Rumford) wanted to test was the motion theory of heat. And on that understanding it is clear that he was a great experimental scientist.

Beside the numerous scientific biographies, an important re-evaluation of Rumford's work has been undertaken by Professor Sanborn C. Brown. In several articles in the *American Journal of Physics*, and in a recent book,[30] Professor Brown has explained the scope and depth of Rumford's lifelong work on theory of heat; in view of this I shall deal rather briefly with this topic.

Rumford's fame rests on his above-mentioned lecture at the Royal Society in 1798.[31] His argument will be given, mostly in his own words, and is as follows:

Being engaged lately in superintending the boring of cannons in the workshops of the military arsenal at Munich, I was struck with the very considerable degree of Heat which a brass gun acquires at a short time in being bored, and with the still more intense Heat (much greater than that of boiling water, as I found by experiment) of the metallic chips separated from it by the borer. The more I meditated on these phaenomena, the more they appeared to me to be curious and interesting. A thorough investigation of them seemed even to bid fair to give a farther insight into the hidden nature of Heat; and to enable us to form some reasonable conjecture respecting the existence, or non-existence, of an *igneous fluid*—a subject on which the opinions of philosophers have in all ages been much divided.

Now Rumford lists the questions that he asked himself. First he wanted to know 'from whence comes the Heat actually produced in the mechanical operation'. One possible answer seemed to be, that the heat was furnished by the metallic chips separated by the borer. Could the 'modern doctrines' explain this?

If this were the case (that the chips supplied the heat) then according to the modern doctrines of latent Heat, and of caloric, the *capacity of Heat* of the parts of the metal, so reduced to chips ought not only to be changed, but the change undergone by them should be sufficiently great to account for *all* the Heat produced.

That is, according to the caloric theory, if now the pieces of metal released their 'combined caloric' (their capacity of heat)—their capacity to retain it in a combined form must have been reduced seriously. This he tested experimentally and found that their capacity of heat had not changed. Thus, a different source for the heat has to be found. But before continuing the Count's argument, let us remember the following. Although his whole line of argument is a theory of 'falsification', admirable as this may seem to us today, one cannot but wonder at the complete absence of an attempt at a theory or of how these phenomena could be explained on an atomic model. True, the magic words 'heat is motion' will appear very soon, but what does this mean in the terms of the corpuscular philosophy of the late eighteenth century? Without a concept of energy, which again implies a theory of conservation of energy, the motion theory of heat has its strength exactly in the same factor in which it has its failure—in ignorance. The motion theory of heat does not supply at this stage quantitative, falsifiable conjectures. (This choice of Popperian Language, which is due to the fact that Rumford seems to have been an ardent Popperian in his procedure to discard the caloric theory, would make every Popperian rejoice; but he does not do the same with his alternative theory.)

Another comment is unavoidable here, though its value is more suggestive than demonstrative. Rumford talks here explicitly of the 'modern doctrine'; he considers the motion theory of heat as the old doctrine, and seemingly credits it with serious scientific status. Could this possibly reflect his social attitudes of preferring in all aspects of life the old to the new, the traditional to the modern? His terminology—'igneous fluid' and 'caloric' used interchangeably—reflects his acquaintance with Lavoisier's *Traité*, probably in the English version of Kerr which appeared in 1790 in Edinburgh.

The experimental proof that latent heat does not change by the creation of the metal chips is rigorous even by our standards, and very ingenious, though I shall not go into detail. At the end of this experiment, Rumford already remarks:

Finding so much reason to conclude that the Heat generated in these experiments, or excited, as I would rather choose to express it, was not furnished *at the expense of the latent Heat or combined caloric* of the metal . . .

The choice of 'excited' clearly indicates the conceptual framework of Rumford's mind at that stage.

After this, he checks for safety's sake whether air did not contribute somehow to the 'excited' heat. During these experiments he calculated with a good degree of exactitude the rate of the production of the heat, which will show numerically

how large a fire must have been, or how much fuel must have been consumed, in order that, in burning equably, it should have produced by combustion the same quantity of Heat, in the same time . . .

as in his experiments. All the calculations and the experiments convincingly prove that indeed 'caloric' is not the explanation of all these phenomena:

And, in reasoning on this subject, we must not forget to consider that most remarkable circumstance, that the source of the Heat generated by friction, in these experiments, appeared evidently to be *inexhaustible*.

It is hardly necessary to add, that anything which any *insulated* body, or system of bodies, can continue to furnish *without limitation*, cannot possibly be a *material substance*; and it appears to me to be extremely difficult, if not quite impossible, to form any distinct idea of anything capable of being excited and communicated in the manner the Heat was excited and communicated in these experiments, except it be MOTION.[32]

(Needless to say, in all these quotations the italics and capital letters are those of Rumford himself!)

This last passage is the most often quoted passage in Rumford, largely because of its last sentence, which is indeed important. Nevertheless, it seems to me that the beginning of it is not less important, since it clearly shows not only that Rumford did not have any conservation ideas in mind, but the very agument by which he proves that heat is not caloric is that it is not conserved! He repeatedly experimented for years and his conclusion was that heat can be produced in inexhaustible quantities by friction, or even better, as he repeated in one of his later papers, by percussion. It was this aspect of the problem which Planck must have meant (though not mentioning any particular historical instance) when he said that the commitment to the belief that no *perpetuum mobile* can be created was no sufficient foundation for a belief in a conservation law. Rather:

Dauerte es doch geraume Zeit, bis man zu der weiteren, hier höechst wesentlichen, Erkenntnis kam, dass jener Satz sich auch *umkehren* lasse, dass es also auch keine Vorrichtung gibt, durch welche sich Arbeit oder lebendige Kraft fortwährend verbrauchen liesse, ohne eine anderweitige als Kompensation zu betrachtende Veränderung.[33]

In this context we should also remember, that although in some of the experiments Rumford used horses to turn the apparatus, and though he counted exactly the number of revolutions, there was never any attempt on his part to calculate a mechanical equivalent of heat, something with which he is very often credited. As to the parameters of the experiment, which seem to us mostly unnecessary today (and that is the reason we pay such great attention to some which seem to us as significant like the number of revolutions of the cylinder), Professor Brown's explanation is very illuminating:

Thus besides being a report of considerable scientific merit, it includes . . . facts which now appear to us irrelevant, such as the date, the weather, the barometric pressure, the relative humidity, the air temperature, and so on, which at that time were not so out of place because Rumford did not know whether these facts were relevant to his experiment or not.[34]

One should consider the data about the horses and the number of revolutions of the cylinder in a given time in this light. In the first chapter of the same book Sanborn Brown also points out that Rumford did not have an energy point of view:

Rumford's own theory of heat was a vibratory one . . . and this hypothesis did not require a conservation of energy point of view so that the whole issue of an energy equivalence was never raised.[35]

This remark does not, however, emphasize the point I made earlier—not only did Rumford not have a conservation of energy view, but his very argument is based on the absence of any conservation. Lest one should think that this was a chance remark on his part, let me quote from a letter Rumford wrote Professor Pictet in Geneva, on November 8, 1797:

The results of my Experiments seem to me to prove to a demonstration that there is no such thing as an *Igneous fluid*, and consequently that Caloric has no real existence. You must not however raise your expectations too high respecting my experiments. Though they were made on a large scale, and conducted with care, there was nothing very new or very remarkable about them; and as to their results they proved only the single fact—that the heat generated by friction is inexhaustible . . .[36]

The truly remarkable fact about the experiments is even more than the interpretation that heat must be motion—that in Rumford's view it is an old and well-known theory!

Masao Watanabe, in an article, 'Count Rumford's First exposition of the Dynamic Aspect of Heat',[37] points out that though the above-cited paper by Rumford is by far the best known, nevertheless, his interest in the motion theory of heat dates back further. Watanabe states that his early 'Experiments upon Gunpowder' in 1781 (on which he began working in 1778) show the direction of his thoughts. This is perfectly true, and also proves that Tyndall's sweeping statement that 'Rumford, in this memoir annihilates the material theory of heat' is very misleading. Rumford worked for more than thirty years on this subject, and whether he held it as a preconceived idea or not, he was too good an experimenter to have 'annihilated' a rival theory on the strength of one experiment! (This same statement by Tyndall is also quoted by Professor Brown, but it is so strong and so typical that I have to repeat it here.) Incidentally, those lectures of Tyndall—'Heat as a Mode of Motion'—where he repeatedly refers to Rumford and Davy, might also be the source of the oft-quoted error that Rumford had an almost clear idea of the mechanical equivalent of heat. Immediately after describing Rumford's experiments, and using expressions like 'horse-power' and 'energy', Tyndall says:

Mr Joule has estimated the amount of mechanical force expended in producing the heat (mentioned by Rumford) and obtained a result which 'is not very widely different' from that which greater knowledge and more refined experiments enabled Mr Joule himself to obtain . . .[38]

Ever since this remark (1869) it has been widely believed that Rumford had meant to do the same calculation. The experiments which Rumford conducted over the years covered a wide range of problems. He proved conclusively that there was no need for a caloric theory in order to explain the transport of heat in a fluid. He showed that the motion of the liquid itself was responsible for the transfer of heat, and what is even more significant, that if the convection currents stopped (the term 'convection current' had originated in Prout's 1834 Bridgewater Treatise), the fluid acted as an insulator. He also showed that even at constant temperature there was internal motion in fluids. Rumford also did important work on the theory of radiation, although, he believed (like his friend Professor Pictet) that there was calorific as well as frigorific radiation. This was very well suited to his theory of heat, and cannot be simply discarded as a simple error, though error it was. By another brilliant argument, he also showed that no absolute zero temperature should exist, again a result which was a direct conclusion from his theory that heat was 'MOTION'.

Humphry Davy: general
We come now to Sir Humphry Davy, doubly interesting for us, because he was not only a working scientist but also a man who held philosophical

opinions upon which he acted. And even more because at all times he was credited with more philosophical opinions than could possibly have been held by one person. In the scientific world of the early nineteenth century[39] he was a dominant figure. His brilliance, wit and good looks, combined with a prolific pen and rhetorical talents all contributed to the fact that he left his mark on most aspects of early nineteenth-century intellectual life. He raised chemistry to a new dignified status in England, a status, which ever since Newton had been accorded only to physics (under the name of Natural Philosophy), or to Moral Philosophy. He turned the Royal Institution into a prospering social centre, and a centre for popular education (though not in those circles for which it had been intended originally by Count Rumford). He travelled widely on the Continent where he spread the fame of English science, a badly needed service after the decline of science in England during the eighteenth century. The French considered Newtonian science as their private heritage, and indeed brought it to new and previously unknown peaks. Davy is also credited with having created for the first time an international atmosphere in science. The story of his going to France while England was at war with that country on a special permit of the French Government, is often cited as evidence. Without going into details, it is enough to follow the furore in the English papers after his acceptance of the French offer and travel permit, and also the letter Davy wrote about the French to Lord Liverpool, First Minister of the Crown. Davy was also the patron and teacher of the great Faraday, important for personal and conceptual history. Thus we have simultaneously to examine Davy's theory of heat, his metaphysical commitment and his philosophy of science.

Davy's often quoted contribution to the dynamical theory of heat (again, we should rather speak of dynamical theory of heat, or simply the theory that heat is a mode of motion, for one can justifiably speak of a 'mechanical theory of heat' only after the work of Clausius) is his paper of 1799, 'An Essay on Heat, Light and the Combinations of Light'. Here Davy states in unequivocal terms that heat is not material, but a form of motion and believes himself to have proven this experimentally. In describing the 'Essay' I will endeavour to show that (a) being an early work of Davy, it is written without proper scientific background, on the grounds of *a priori* commitment to the theory that heat is motion, taken from Bacon, Locke and the physiologists; (b) that his experimental proofs are completely unreliable; (c) that with the passage of time, not only did he himself express doubts about the value of his early contribution, but that the more science he knew, and as his philosophy of science took shape, he became more and more lenient towards the caloric theory, until in the 1820s he no longer considered the question settled one way or another.

Humphry Davy: his 'motion theory'

At the age of nineteen Davy was appointed by Dr Beddoes to be the superintendent of the Pneumatic Institution in Clifton. Dr Beddoes became interested in him after having heard of his discovery, while still at Penzance, that

oxygen is essential to the existence of these classes of animals (fishes and zoophytes), that they breathe air contained in water, and like the higher class of land animals, convert the oxygen by addition of carbon, into carbonic acid gas.[40]

This 'Essay' was his first scientific paper, written at the age of nineteen without any fundamental scientific training. It grew out of his work at the Institute, where the task allotted to him was

to make trial of different gases for the purpose of ascertaining their medicinal effect in various diseases.[41]

In the introduction to the 'Essay', Davy emphasizes that chemistry so far had not contributed much to physiology and medicine, although the hopes had been great.

Even before analysing the work itself, it is needless to say that he did not have the experimental background from which to draw the inductive conclusions about the nature of heat, with which he is often credited. Even a man of Sir Humphry Davy's scientific calibre cannot be said to have drawn a far-reaching inductive conclusion from two experiments, even if they would be creditable. His brother reports that his first interests were metaphysical and that he read carefully and left notes about his readings in

the distinguished metaphysicians of modern times, as Locke, Hartley, Bishop Berkeley, Hume, Helvetius, Condorcet, Reid, and his followers, who are designated by the general title of 'Scotch metaphysicians'; and he appears to have had some acquaintance with the doctrines of Kant and the Transcendentalists.[42]

From various extracts and comments it would appear that he was much occupied with Locke (hence at least one source of his theory of heat) and also Condorcet. As to Kant we shall see later, but Davy did not read German, and the *Metaphysische Anfangsgründe der Naturwissenschaft*—which, in the opinion of Professor Pearce Williams[43] influenced Davy through the intermediary of Coleridge—did not exist in English. (This is mentioned by Pearce Williams very clearly, but he relies on Coleridge's influence; about this more later.) Some remarks in the introductory chapter on the theory of perception and 'trains of connected sensations called facts' sound as if taken straight out of Locke. Whether one can assume, as Merz[44] does, that his interest was basically medical, I do not know, but I am inclined to doubt it.

Davy's first aim is to show that the answer to Lavoisier's query:

la lumière, est-elle une modification du calorique, ou bien le calorique est-il une modification de la lumière?[45]

is that neither possibility is true. An experiment is described which is supposed to prove that 'light is not a modification, or an effect of heat'. It may throw some light on the experimental value of the ice-rubbing experiments, in which Davy believed himself to have shown that particles of iron could be brought to melting point without giving out light! After this he concludes that light is a matter of 'peculiar kind' while the next experiments will prove that 'heat, or the power of repulsion is not matter'. This remark brings us nearer to the scientific background of Davy in these early days of his career. His theory of attraction and repulsion is basically the current form of the Newtonian corpuscular theory. His background in chemistry is Lavoisier's *Traité*. Here, in the wake of Lavoisier and Black, and probably influenced by Condorcet, he also tries his hand at creating a new nomenclature. He read Lavoisier's 'caloric' as if it dealt only with the problem of heat and not with light too (Lavoisier says clearly that release of caloric generates both heat and light), and thus, having proved that heat is immaterial he discards the theory that oxygen is a compound of oxygen and caloric. He then proposes that it is in reality a compound of oxygen and light; for this reason he suggests that it should be called 'phos-oxygen'. Davy, in his later work, was never a creator of new concepts, and no scientific term can be traced to him—his creativity was of a different kind. In view of this, I doubt if he actually said that 'no affection of singularity induces me to use a new nomenclature'.[46] His theory of light-perception is an interesting and critical version of the Newtonian optical theory and the Lockean theory of perception. One of the most interesting passages is Davy's description, as to the kind of particles of which light consists:

Light is a body in a peculiar state of existence. Its particles are so amazingly minute, that they are very little affected by gravitation; unaltered through the pores of diaphanous bodies. They move through space with a velocity almost inconceivable, and communicate no perceptible mechanical motion to the smallest perceptible particles of matter. From the peculiar velocity of light we estimate its quantity of repulsive motion. The influence of the attraction of gravitation on light is very small, as is evident from its not apparently gravitating towards the sun or the earth.[47]

I quoted this passage because it gives us the first inkling of the real Davy, and a taste of where his real power lies: it is in the balanced view of every scientific theory, with very much a common-sense combination of hypotheses and experimental facts. The formulation is very careful and uncommitted—like the later Davy, and not like most of the 'Essay' of which this passage is a part. It also shows that his philosophical back-ground, as far as it was influential, was pointed much more towards experimental science than towards any far-reaching metaphysical commit-ments. As far as scientific temper goes, he was more of a Maxwell than a Faraday, which only proves that the mathematical genius of Maxwell is only one of the possible forms into which his kind of scientific tempera-ment can develop.

Humphry Davy: the 'ice-rubbing experiment'
The famous ice experiments on which Davy's theory of heat is founded is so lucidly described that it will be best to quote it verbatim:

I procured two parallelopipedons of ice, of the temperature of 29°, six inches long, two wide, and two-thirds of an inch thick; they were fastened by wires to two bars of iron. By a peculiar mechanism, their surfaces were placed in contact, and kept in a continued and violent friction for some minutes. They were almost entirely converted into water, which water was collected, and its temperature ascertained to be 35°, after remaining in an atmosphere of a lower temperature for some minutes. The fusion took place only at the plane of contact of the two pieces of ice, and no bodies were in friction but ice. From this experiment it is evident that ice by friction is converted into water, and according to the supposition its capacity is diminished; but it is a well-known fact, that the capacity of water for heat is much greater than that of ice; and ice must have an absolute quantity of heat added to it, before it can be converted into water. Friction consequently does not diminish the capacities of bodies for heat.

From this experiment it is likewise evident, that the increase of temperature consequent on friction cannot arise from the decomposition of the oxygen gas in contact, for ice has no attraction for oxygen. Since the increase of temperature consequent on friction cannot arise from the diminution of capacity, or oxydation of the acting bodies, the only remaining supposition is, that arises from an absolute quantity of heat added to them, which heat must be attracted from the bodies in contact. Then friction must induce some change in bodies, enabling them to attract heat from the bodies in contact.[48]

The inconclusive character of this experiment from a scientific point of view has been demonstrated beyond any doubt by Professor E. N. da C. Andrade. Again, for the sake of clarity, we shall quote:

Now any physicist who contemplates repeating the experiment, will, I think, at once be struck with the difficulty, if not impossibility of carrying it out in such a way as to produce anything in the nature of a convincing result. If the ice is covered with a film of water, the friction is so small that scarcely any work is done, while if it is really dry it is liable to stick. In any case, to make the frictional heat appreci-able, it is necessary to have a normal force holding the two surfaces together, and then one gets the well-known lowering of freezing point and consequent melting if the surroundings are at the ice point with all the possible dangers of regelation at the edges. Again, the amount of work required to melt 1 gm. of ice is very large:

the criterion is an extraordinarily insensitive one. All these difficulties are, perhaps, sufficiently summarised in the fact that nobody, apparently, has ever tried to repeat the experiment, and I, for one, would not care to undertake it. . . . The whole experiment is fantastic. This is said in no disrespect to Davy: how could one expect an untrained boy in 1799 to carry out an experiment which even to-day would tax an experienced physicist, to say the least? No doubt the whole effect observed by Davy was due to conduction.[49]

Without getting into a character study of Davy, it is well-known that he was not without ambition. It is also known that Count Rumford invited him to the Royal Institution because of this 'Essay', and especially because he was impressed by Davy's theory of heat. Naturally the question arises whether Davy knew about Rumford's theory. Most of Rumford's papers on these subjects were published later than 1799, but not all, and besides, his ideas were often talked about. I shall not try to answer this question here; let me only point out that Davy was aware of this, and in a later addendum to his 'Essays' wrote:

The experiments on the generation of heat were made long before the publication of Count Rumford's ingenious paper on the heat produced by friction. His experiments alone go far to prove the non-existence of caloric, and when compared with the second and third experiments in my essay, will, I should conceive, leave no doubts on the mind of the impartial and philosophic reasoner.[50]

I wonder why Davy does say that. True, Rumford's very important reviews on the subject appeared in French in two books: *Mémoire sur la Chaleur* in 1804, and the *Récherches sur la Chaleur développée dans la Combustion et dans la Condensation des Vapeurs* in 1813, and Davy might have meant the first of these. But Rumford's very important 'Experimental Inquiry Concerning the Source of Heat which is excited by Friction' was read before the Royal Society on 25 January 1798, and published in six places, namely:

> *Philosophical Transactions* lxxxviii (1798), 80–102
> *Bibliothèque Britannique* viii (1798), 3–34
> Nicholson's quarto *Journal* i (1798), 459–68, 515–18
> Gilbert's *Annalen der Physik* iv (1798), 257–81, 377–99
> Scherer's *Journal der Chemie* i (1798), 9–31
> Voigt's *Magazin* i (1798), 94–106[51]

Is it possible that Davy did not know about this paper?

Humphry Davy: his development

What I tried to show above was no attempt to belittle Sir Humphry Davy. He does not need anybody to defend him; yet, let me say it clearly: he was one of the great scientists of his age and his contributions to chemistry and electricity formed important stepping-stones in the histories of these sciences. But to treat his first weak attempts in science as all-important links in later conceptual developments, just because he accidentally turned out to have been right, is a gross error in judgement. How prejudiced posterity can become is clearly shown by the fact that nobody ever took his first experiment (about the melting particles of iron that do not emit heat) seriously, while his ice rubbing experiments are quoted everywhere more than twenty years after Andrade's paper. Interestingly, in 1881, J. B. Stallo in his *The Concepts and Theories of Modern Physics* remarked:

I am aware of course of the anticipations of the modern theory of heat by Bacon, Locke, Rumford, Sir Humphry Davy etc. but their announcement however clear, that heat is but a 'mode of action' received but little attention. . . .[52]

The more science Davy knew the more careful in formulation and more balanced in judgement he became. In 1810, among the extracts published by his brother, we find the following statement:

I do not think we have at present any means of deciding upon this question of the nature of heat; and its effects may be studied,—and it may be employed as an instrument of experiment, without the necessity of adopting any hypothetical views respecting its cause. It is my intention merely to give a caution with respect to the adoption of the chemical solution,—which is by far the most generally received. Indeed, the matter of heat or caloric is sometimes talked of with the same confidence as water, or any common ponderable bodies. It is of great importance to the progress of science, that facts should be separated from what is imagined; that the nature of our knowledge, and the grounds of our opinions, should be strictly defined. The truly philosophical inquirer into nature will not consider it as a disgrace, that he is unable to explain every thing; he will wait, and labour with hope, tempered by humility—for the progress of discovery;—and he will feel that truth is more promoted by the minute and accurate examination of a few objects, than by any premature attempts at grand and universal theories.[53]

In the chapter 'On Powers and Properties of Matter and the General Laws of Chemistry' of his *Elements of Chemical Philosophy*, published in 1812, there is a long discussion on the nature of heat. He introduces paragraph 16 by explaining that the 'subtile' fluid named 'matter of heat' or 'caloric' has been introduced to explain 'calorific repulsion', and also

Many of the phenomena admit of a happy explanation on this idea, such as the cold produced during the conversion of solids into fluids or gases, and the increase of temperature connected with the condensation of gases and fluids.

But production of heat by friction and percussion cannot be explained by the caloric theory, and some of the chemical changes also run into difficulties. After this Davy repeats Rumford's argument:

By a moderate degree of friction, as it would appear from Rumford's experiments the same piece of metal may be kept hot for any length of time; so that if heat be pressed out (on the caloric theory heat is pressed out of the metal when beaten, and this makes it hot), the quantity must be inexhaustible!

Again the same argument: the very fact that it is not conserved, but seems to be inexhaustible is the argument that heat could not be matter! A few lines later Davy summarizes that

the immediate cause of the phenomenon of heat, as Lavoisier long ago stated, is motion, and the laws of communication are precisely the same as the laws of the communication of motion.

Lavoisier's is a good name to use for any purpose! One would expect that this is a summary of the argument. But this turns out not to be the case. A few paragraphs later we find:

Very delicate experiments have been made to show that bodies when heated do not increase in weight. This, as far as it goes, is an evidence against a specific subtile elastic fluid producing calorific expansion; but it cannot be considered as decisive, on account of the imperfection of our instruments.

What conveys Davy's approach more than anything is the last paragraph of article 16, where he shows that this does not matter to him very much. We shall see in the next sections that one can find in Davy's works as many quotations as one wishes to make him either an instrumentalist-positivist or a Kant-and Schelling-influenced 'Naturphilosoph'. In any case his approach to the theory of heat is neatly summarized in the following, which also proves conclusively that no conservation ideas could have been at the back of his mind:

The laws of communication of heat, and the philosophy of its effects, are independent of this speculative question, which has been mentioned in this place merely with the view of guarding students against the adoption of the doctrine of a specific fluid of heat as a part of the philosophical principles of chemistry, and to show that as yet we have no decided evidence on the subject.[54]

Finally, in 1820, when Davy was no longer very active in creative scientific work, his Presidential Address to the Royal Society was 'On the Present State of the Royal Society and on the Progress and Prospects of Science'. Here he said:

The subject of heat, so nearly allied to that of light, has lately afforded a rich harvest of discovery; yet it is fertile in unexplored phenomena. The question of the materiality of heat will probably be solved at the same time as that of the undulatory hypothesis of light, if, indeed, the human mind should ever be capable of understanding the causes of these mysterious phenomena.[55]

One should not be surprised by these changes and undecided issues.

Davy was a very gentlemanly thinker. No intellectual excesses and no excited commitments characterize either his personal or his scientific life. As it is exactly this philosophical quality which his brother admired so much, we have the whole spectrum of his opinions, although the extracts which have been published in the *Collected Works* and in the *Fragmentary Remains* have all been selected by Dr John Davy from the numerous unpublished notebooks and diaries. Among the same extracts we find one in 1800:

Consistency in regard to opinions is the slow poison of intellectual life, the destroyer of its vividness and energy.

Davy's use of the word 'energy'

Before dealing with the philosophical side of the work of Sir Humphry Davy, I wish to remark here on the very interesting usage of the word 'energy' in his writings. It is well known that the word was used not infrequently in the late eighteenth and early nineteenth centuries in a literary, vague sense, not very far from the non-technical use of the Greek 'energeia'. To clarify, Aristotle uses the word 'energeia' in two different senses—in the literary, non-specific sense of 'mental energies' and in the technical sense as against 'dynamis'—the famous philosophical twins of 'potentiality' versus 'actuality' which had such a long history in scholasticism and medieval science. It is also well known that it was Dr Thomas Young, Davy's contemporary, who first introduced the term as a technical term, to rename what Leibniz called '*vis viva*'. Young introduced the term in his 'Lectures on Natural Philosophy', held at the Royal Institution in 1802–3 (at a time when Davy was already there) and then published in 1807. Why is it that Davy never took up the technical use of a word which he frequently used, indeed as we shall see used much more frequently than one could expect if it were done casually or unintentionally? The relations between these two great English scientists, whose lives ran parallel in so many respects, are very intriguing and not at all documented. Why do the two men refer to each other's work only on the rarest occasions and even then only in insignificant passages? Davy held to the corpuscular theory of light while Young's contribution to the undulatory theory needs no comment. They worked together at the Royal Institution for more than two years; in 1803 Young was refused by the managers some request, after which he resigned. Davy was the foremost popular lecturer of his day, while there were many complaints that Young's lectures were incomprehensible to the audience and even boring. One cannot get rid of the suspicion that Davy, if not instrumental in Young's resignation, certainly did not do anything to prevent it. In all fairness to Davy one has immediately to add that even according to Dr

Young's biographers (Drs Peacock and Alexander Wood), Young was a complete failure as a lecturer. Also, it is well known that the Royal Institution was dependent on the quality of its lecturers and their popularity, to obtain private subscriptions and popular goodwill. Nevertheless, in later years they had some quasi-scientific disagreements, mainly on the proper way of unrolling the Herculaneum papyri. Among the letters of Young there is one to his friend Gurney in 1820, where he says:

Davy has been giving dinners as P.R.S.—not better certainly nor pleasanter than he used to give before.[56]

As we shall see now Davy continued to use 'energy' in its literary sense, and even when he used it in a scientific context, and not in poems, his use was intentionally vague.

Davy used to write poems from a very early age. There is considerable discrepancy of opinions on the literary value of these: the opinions diverge from 'superb' to downright 'mediocre'. In any case in his time serious poets did read his poetry and often spoke about it in very laudatory terms—Southey, Wordsworth and Coleridge among others. In almost every one of these poems the word 'energy' appears several times.

> Then, as awakening from a dream of pain,
> With joy its mortal feelings to reign;
> Yet all its living essence to retain,
> The undying energy of strength?[57]

. . . those deep intense feelings, which man sometimes perceives in the bosom of Nature and Deity, are presentiments of a more sublime and energetic state of existence. . . .[58]

. . . still all my passions, all my energies are alive . . .[59]

. . . I feel within me new energies . . .[60]

> The mystic laws from whose high energy
> The moving atoms, in eternal change,
> Still rise in animation.[61]

The energy of the understanding employed upon the development of the truths of nature has a calm and quiet progress.[62]

His mind seemed to become more active and more energetic in proportion as his bodily strength decayed.[63]

> While superstition rules the vulgar soul,
> Forbids the energies of man to rise,
> Raised far above her low, her mean control,
> Aspiring genius seeks her native skies.[64]

The force of this argument is only in the sheer quantity of the instance where this use of 'energy' occurs; I could not collect all of them for fear of boredom, but when one reads through the *Collected Works* one is struck by the repeated appearance of the term in this context. Let us now look at the more scientific contexts of our term. Davy introduced a term 'electrical energies', and one is tempted to assign to this some kind of physical meaning. But in vain.

Whatever be the fate of the opinion that electrical energy and chemical attraction are one and the same power, the facts of the dependence of the chemical arrangements of matter upon electrical functions will be permanent . . . [65]

The energy of combination belongs, in the highest degree, to simple bodies, and it almost disappears in very complex substances. [66]

In these two quotations we might have been led to detect some sort of physical meaning; but what sort of 'power' could a 'chemical attraction' which is 'electrical energy' be? Especially if it is only in simple bodies? Could it be a concept of valency? We approach the answer when we examine further instances. Here we begin to see that Davy talks of 'natural energies' and generally of two different energies:

In the great phenomena of the solar system, the various and harmonious mutations of the heavenly bodies are accounted for, by the supposition of two energies operating upon matter, gravitation and the projectile force, and the laws of these have been submitted to calculation by the power and sagacity of the human genius. [67]

Bodies contracting, suffering a diminution of volume by cooling, on the contrary, are positive—and, as it were, communicate electricity to the surrounding bodies, whatever may be their mutual energies. [68]

And in a few paragraphs before that:

In the operation of the common electrical machines, the electrical fluid has been generally conceived as pressed out of the glass by influence of friction. The theory on the principle of natural electrical energy, will certainly be less complicated and mechanical. [69]

But if we tend now to call the two kinds of electricity simply the two natural electrical energies, how shall we understand the following?

The materials of the air, of the solid surface, and of the ocean, differ in electrical energy. [70]

Among the *Miscellaneous Papers* there is one titled 'On the Relations between the Electrical Energies of Bodies, and their Chemical Affinities'. This begins with an explanation which again shows us a new feature of Davy's 'electrical natural energies':

As the chemical attraction between two bodies seems to be destroyed by giving one of them an electrical state different from that which it naturally possesses; that is by bringing it artificially into a state similar to the other so it may be increased by exalting its natural energy.[71]

In my opinion, Davy is using a deliberately vague or non-specified term, in order to be able to talk of things which are not yet clearly understood, without further confusing the issue. Let us remember, that unlike those who talked of 'forces' instead of talking of 'energy' or some other physical term, in the 1850s Davy did not confuse his readers by using the term 'energy'. In his time there was no other connotation of the word in use than the literary one, and it suited the purpose very well. That indeed this was Davy's purpose in the choice of the term is supported by one of his rare methodological remarks on scientific nomenclature:

In the nomenclature of electricity, commonly adopted in this country, positive, or plus, is synonymous with vitreous, negative or minus, with resinous electricity. The words positive and negative, plus and minus, are sanctioned by the authority of Franklin, who has accounted for the phenomena by the supposition of one fluid only in excess in bodies possessing the vitreous electricity, in deficiency in those possessing the resinous electricity.

It is unfortunate for the diffusion of science, that any terms should be in common use which imply hypothetical ideas; but, in the present early stage of the investigation, it is better to adopt even an imperfect language than to form a new one. The denominations, as I shall use them, you will have the goodness to consider as merely signifying the unknown causes of a certain series of facts.[72]

As in all his other philosophical opinions, here too Sir Humphry Davy covers the whole range of diverse uses. I have purposely left two quotations to the end. After the foregoing, I think my point will have been carried, if we now look at one passage where 'electrical energy' is almost synonymous with 'kind of electricity' while in the second, Davy uses 'energy' in its vaguest sense:

If the electrical energies of bodies are examined, oxygen, and all bodies that contain a considerable proportion of oxygen, appears to be *negative*; hydrogen, the metals, and all combustible bodies, *positive*.[73]

To be able to understand the permanency or the changes of forms of bodies, the series of events in the history of Nature, and in the operations of art, it is necessary, as I mentioned in the Introductory Lecture, to consider the active powers belonging to matter, and the laws of their operation.

By active powers are understood those powers which cannot be separated from the bodies which they affect, and which produce the motions of their particles—such as the expansive energy of the power of repulsion which produces heat, and attraction in its different modifications—as gravitative, chemical, or as electrical attraction.[74]

Davy's philosophy

Coming now to Sir Humphry Davy's philosophy, or rather to his philosophy of science, we can easily follow the maturing of his mind and his gradual development into a paradigm of common sense and balanced wisdom. At the beginning of his career, he was influenced by the 'Scottish Metaphysicians' as we have seen above. His brother also speaks of having found among one of Davy's early notebooks the following remark:

(I) began the pursuit of chemistry by speculations and theories; . . . more mature reflection convinced me of my errors of the limitations of our powers, and the dangers of false generalizations.[75]

In the 'sketch of his life' which John Davy inserted into Volume I of the *Collected Works*, we find the same passage but here he explains:

This I find written in darker ink between the lines of a note-book kept in 1799, consisting chiefly of speculative views concerning the connexion of life and chemical action, or of physiology and chemistry.

John Davy thinks that this is explained by his brother's having become acquainted with the Baconian maxim 'homo Naturae minister, et interpres tantum facit et intelligit quantum de Naturae ordine re vel mente observaverit: nec amplius sit, aut potest'. In many case these were written shortly after the publication of his essays. Davy even goes on to say:

I was perhaps wrong in publishing with such haste a new theory of chemistry. My mind was ardent and enthusiastic. I believed that I had discovered the truth. Since that time my knowledge of facts increased—since that time I have become more sceptical.

In later years Davy became acquainted with Bacon's and Newton's philosophy of science and the depth of his reading of their philosophy much surpassed the one-sided, prejudiced versions that early nineteenth-century scientists used to give. Davy was a careful experimenter and a bold conjecturer, alert to new theories, prone to change his opinion on any subject on sufficient new evidence, not committed to any extreme view of science or the method of science. Indeed, so flexible was his mind, and so versatile his announcements, that every historian claims him for his own type of hero. This certainly is an irony of fate, since Davy is the least suitable type of intellect to become anybody's hero. He is not a positivist, not an operationalist, not a Scottish metaphysician, and not a 'Naturphilosoph'. This is not a list of philosophical schools for the sake of amusing ourselves, for Davy was called every one of these by some historian or other. The 'inductivist' Whittaker deletes passages with conjectures and unfounded hypotheses by the dozen and claims Davy to be one of his heroes. Most of the historians see him as an ardent supporter of the mechanical theory of heat and quote it as a beautifully established inductive proof. The value of that we have already seen; but even more: it is generally supposed that not only was he undecided between the caloric and the motion theories of heat, but at various times he tried to reformulate the phlogiston theory in an acceptable way, so as to suit the experimental evidence which had been accumulated in the meantime.[76]

J. Z. Fullmer, in an article on Davy's poetry, goes to such an extreme of misjudgement that I have to quote her:

One of the most fruitful of scientific generalizations has been the first law of thermodynamics, which concerns just these notions of permanence and change and eternality. Although Davy's scientific work contains no precise statement of the laws of thermodynamics the poems show that he held the fundamental point of view necessary for such a statement. It is lodged so deeply in his poetry that it also must have been part of the *Weltansicht* against which he conducted his experiments.[77]

The poem in which Fullmer seems to discover this is in the 'Fair Head' fragment in Davy's 'Mucrish and Arokil':

> Long shalt thou rest unalter'd mid the wreck
> Of all the mightiness of human works;
> For not the lighting, nor the whirlwind's force,
> Nor all the waves of ocean shall prevail
> Against thy giant strength, and thou shalt stand
> Till the Almighty voice which bade thee rise
> Shall bid thee fall.[78]

The nearest to a conservation idea in Davy's whole output can be found in one poem, again quoted by Fullmer to prove her theory. How farfetched this interpretation is will be seen immediately if we look at the verse itself:

> Nothing is lost; the ethereal fire
> Which from the farthest star descends,
> Through the immensity of space
> Its course by worlds attracted bends,
> To reach the earth.[79]

Not only does Fullmer[80] imagine to have discovered here the law of conservation of energy, but even the general theory of relativity, with the gravitational attraction of light. We have already seen above Davy's approach to this problem; for those who held to the corpuscular theory of light in the late eighteenth century this was a usually accepted conclusion.

Pearce Williams, whose penetrating study of Faraday will form the foundation of my analysis of Faraday, gives a picture of Sir Humphry Davy's philosophy which I cannot but oppose. Professor Williams traces the Kantian influence on Faraday, and more generally the influence of the German 'Naturphilosphie' on English science, and he proves, conclusively I think, that Coleridge was the vehicle which brought this philosophy to England. However, Professor Williams finds that the connecting link between Faraday and Coleridge was Sir Humphry Davy. Now Davy was an intimate friend of Coleridge, and the teacher and mentor of Faraday; all this seems to work out nicely, yet it seems to me that it is not true. It would be a worthwhile study to find out who constitutes the missing link (or perhaps one should leave more place to individual temperament and character than we tend to do in our conceptual histories chasing possible influences) but it was not Humphry Davy. I will attempt to show that Davy's philosophy is far from related to Kantianism or indeed to any metaphysical school. After that I shall examine the evidence given by Professor Williams as to the Coleridge–Davy relationship. Let me collect at random some of Davy's methodological statements, and arrange them according to the year when they were published or jotted down. We shall see that there is ample evidence both for Pearce Williams' interpretation, and for Robert Siegfried's. The latter writes:

But Davy's objection to the atomic theory was on philosophic grounds. The operational aspects Davy recognized correctly as independent of the theoretical, . . . Davy thought all theories were dangerous. . . . Davy's knowledge of the facts continued to increase until his death, but that increase only more deeply entrenched his scepticism.

1800: If, however, we relinquish wholly speculation and the pursuit of analogies, we do not at all fulfill the end of philosophy:—the most sublime end of the sciences is that of discovering the laws of nature. But our hypotheses alone should be formed with rapidity and applied with ease, and eternally varied; they should be the instruments of thought,—the secret amusements of the mind.[81]

Experiments, even apparently the most trifling, can hardly fail to be useful;—so likewise all insulated systems and partial theories.[82]

1809 or 1810: The only use of an hypothesis is, that it should lead to experiments; that it should be a guide to facts. In this application conjectures are always of use.

Hypothesis should be considered merely an intellectual instrument of discovery, which at any time may be relinquished for a better instrument. It should never be spoken of as truth; its highest praise is verisimility.[83]

. . . it must not be supposed, however, that I am arguing generally against conjectural inferences, or attempting to prove that the imagination ought to be passive in physical research. This would be giving up a noble instrument of discovery.[84]

Nothing is so fatal to the progress of the human mind as to suppose that our views of science are ultimate; that there are no mysteries in nature; that our triumphs are complete, and that there are no new worlds to conquer.[85]

1811: The Huttonian doctrine, considered as an hypothesis, has many advantages over all the other views; it offers probable explanations of many more phenomena, and presents fewer difficulties; difficulties however, it still has, and they must be removed before it can be considered as a genuine theory.[86]

Before commenting on these passages and on some further ones, I wish to raise two points which will not be dealt with at length. There are numerous methodological comments on the interesting scientific usage of analogies in Davy's works. It would need a special study to decide whether this important aspect of his scientific method supports his views on theories as instruments, or if it is rather a deep commitment on his part to a belief in the uniformity of nature. In all probability, as I read Sir Humphry Davy, it would be something in between. The other rich source of quotations which has not been used here is his last philosophical work, or as he called it his Testament, *On the Consolations of Philosophy*, written as a sick and broken man in the last year of his life. That work is full of brilliant suggestions, conjectures and hypotheses, which however were not formulated as guides to experiments, or as mere 'instruments of the mind'. Again, as these ideas were not written down in good health, and not at a time while he was active in scientific research, I did not want to draw from them far-reaching conclusions.

Davy's clearest statement on hypotheses is his 1823 paper 'On a New Phenomenon of Electro-Magnetism'. Here we find:

On a subject so obscure as electro-magnetism, and connected by analogies more or less distinct with the doctrines of heat, light, electricity, and chemical attraction, it is not difficult to frame *hypotheses*; but the science is in a state too near its infancy to expect the development of any satisfactory *theory*.[87]

John Davy remarks in a footnote:

The above distinction between *hypothesis* and *theory* was one of the strongest features of the author's philosophical method,—using the former term in the sense of supposition or opinion, the latter in that of a generalization of facts.[88]

Well, how Newtonian can one be? But to make it even clearer, in his 1820 Presidential Address to the Royal Society, Davy says:

I trust in all our researches we shall be guided by that spirit of philosophy, wakened by our great masters, Bacon and Newton, that sober and cautious method of inductive reasoning, which is the term of truth and of permanency in all the sciences . . . that our philosophers will attach no importance to hypotheses, except as leading to the research after facts, so as to be able to discard or adopt them at pleasure, treating them rather as part of the scaffolding of the building of science, than as belonging either to its foundations, materials, or ornaments.[89]

Thus Davy was neither the 'romantic' type of scientist (Ostwald actually called Davy a 'romantic') committed to sweeping metaphysical principles about the totality of phenomena in nature, nor that caricature of a scientist which some claim Bacon to have advocated. Indeed, he appears to have been a sober-headed critic of science; and some of his works could have been written by Duhem, or even by Popper. The author is interested in the question, whether this sort of common-sense approach is not the death of all great scientific discoveries. Naturally, one cannot say for a specific scientist, that had he been different he would have been a great scientist. However, it is possible patiently to examine how many great creative scientists (using Herivel's or any other criterion) did think and work in such level-headed, uncommitted, critical terms. I think that we would find very few indeed.

Returning to the theory that Davy was influenced by Kant via Coleridge, we have to examine the relations between the two men. Coleridge is not mentioned anywhere in the methodological or philosophical comments of Davy. Kant is mentioned only briefly in a lecture on Electro-chemical Science in 1810, in the following manner:

His [talking of Ritter] errors as a theorist seem to be derived from his indulgence in the peculiar literary taste of his country, where the metaphysical dogmas of Kant, which as far as I can learn are pseudo-platonism, are preferred before the doctrines of Bacon, Locke, and of Hartley—excellence and knowledge being rather sought for in the infant than in the adult state of his mind.[90]

This in itself does not disprove yet that Davy was committed to Kant's views, but nevertheless it should be kept in mind. It is certainly true that Davy and Coleridge were intimate friends, and also there is no doubt whatsoever that Coleridge was an ardent Kantian. This is convincingly discussed in Professor Williams' book,[91] and also in Coleridge's *Philosophical Lectures* which are, I think, the best introduction to Coleridge's philosophy. These were lectures given in 1818–19, were lost and were published for the first time by Kathleen Coburn in 1949.[92] Exactly as Davy does not mention Coleridge anywhere in his works, Coleridge does not mention Davy, although he was very interested in the philosophical implications of chemistry. Even if one would explain that at this time the two were already estranged (as Miss Coburn indicates in the notes to the *Lectures*)[93] yet in the first years of the century Davy invited Coleridge to lecture at the Royal Institution which he indeed undertook in 1808. The explanation seems to me to be something quite different. This friendship was simply not of a philosophical character. The two met in Clifton, while Davy was at the Pneumatical Institute, and from then until late 1810 kept up a correspondence. John Davy finds it very important to repeat again and again how intimate this friendship was. He also tells that unfortunately the bulk of Davy's letters to Coleridge got lost, but he reproduces loyally most of Coleridge's letters to Davy. The interesting

G

point is that there is nothing worth quoting in these letters. In twenty years of correspondence not a single letter mentions any philosophical topic. The letters are full of friendly remarks, narratives on Coleridge's travels and the people he met; sometimes Davy's poetry is discussed with Coleridge's well-meant criticism or on the contrary real admiration for Davy's poetry. We even find scientific questions where Coleridge wants to be taught facts of chemistry, but nowhere any word on metaphysical principles. I do not wish to cast serious doubt on the depth of the friendship (though I cannot quite understand the remark which, according to John Davy, Coleridge had written on the back of a letter from Humphry Davy: 'This from Davy, the great chemist').[94] However, I do not believe that it involved any intellectual exchange of ideas or an attempt on Coleridge's part to exert philosophical influence on Davy. There is certainly no instance in Davy's writings where he mentions the great theory of forces which in Professor Williams' opinion 'he openly adopted'.

There is one early aphorism of Davy's among his extracts which, in my opinion, characterizes his whole system of thought; it is the only one to which he consistently adhered in his thirty years long scientific career, and which would not have let him adopt vaguely formulated, sweeping metaphysical generalizations, especially the German theory of forces; perhaps to the detriment of his science. It is:

Philosophy is simple and intelligible. We owe confused systems to men of vague and obscure ideas.[95]

Carnot and the Second Law of Thermodynamics

There is little doubt that Sadi Carnot discovered the Second Law of Thermodynamics, introduced the cyclical process which we call the Carnot cycle, and recognized the importance of reversible processes in general. Furthermore, his work is beyond the scope of this study. However, a long controversy has been continuing for the last fifteen years between various physicists–historians, regarding the historical and logical place of the First Law of Thermodynamics in the formulation of the Second Law. This is intimately connected with our study. I intend to skip the 'internal history' (as Koenig calls it) of the second law, but will briefly review the controversy and take sides.

Sadi Carnot published his epoch-making 'Réflexions sur La Puissance Motrice du Feu' in Paris in 1824. In this paper (written for a popular audience, for Carnot does not use any mathematics except in footnotes), besides the term 'chaleur' and 'feu' we find the term 'calorique'. With these concepts as his tools of thought he describes the problem facing him, the rational and experimental solutions involved, and comes up with his theorem, which is one of the numerous equivalent formulations of the second law; it is given below in Koenig's paraphrase to his 'History of Science and the Second Law of Thermodynamics':

Call a heat engine *simple* if its thermal interaction with its surroundings consists only in the absorption of heat from a reservoir at a fixed temperature, and the rejection of heat to another reservoir at another fixed temperature; and call any simple engine which is reversible a *Carnot engine*. Then Carnot's theorem is the following. (a) All Carnot engines between the same two temperatures have the same efficiency; (b) if the efficiencies of two simple engines between given temperatures are equal, and one of these is a Carnot engine, then so is the other; (c) the efficiency of an irreversible simple engine between two temperatures is less than that of a Carnot engine between those two temperatures.[96]

The model with which Carnot is working is clearly the caloric theory, and though he makes frequent comparisons between his engine's working between two temperatures, and the fall of a body from one height to another, no clear indication of any conservation is evident. In view of this, it seems undisputable that his 'calorique' is identical with Lavoisier's 'caloric'. Clapeyron, in his 'Memoir on the Motive Power of Heat', attributes to Carnot more than Carnot ever said, but whatever Clapeyron attributes to Carnot is much less than what later historians of science thought that Clapeyron attributed to Carnot. All Clapeyron says is:

The idea which serves as a basis of his [Carnot's] researches seems to me to be both fertile and beyond question; his demonstrations are founded on *the absurdity of the possibility of creating motive power or heat out of nothing*.[97]

If there is one central idea of Carnot's work it would seem to me to be rather 'wherever there exists a difference of temperature, motive power can be produced'. It is also in this connection, dealing with the two different temperatures that Carnot discloses in the clearest way that he thinks in 'caloric' terms.

The production of motive power is then due in steam engines not to an actual consumption of *caloric*, but to its *transportation from a warm body to a cold body*. . . . According to this principle, the production of heat is not sufficient to give birth to the impelling power; it is necessary that there should also be cold.

The very idea of 'equilibrium of caloric' with which all the above-quoted passages are associated is at this stage not a concept from the motion theory of heat.

The historical complication began when in the 1911 edition of *Encyclopaedia Britannica*—the article 'Heat' by H. Callendar suggested that Carnot's 'calorique' is actually the entity which later became known as 'entropy' (so called by Clausius), and that Carnot's model for heat was the motion-theory and not the caloric theory. This criticism was later taken up by several distinguished physicists like J. N. Bronsted,[98] K. Schreiber and L. Brillouin and recently by Professor Victor K. La Mer. La Mer's paper appeared in 1955 in the *American Journal of Physics*,[99] and was conclusively answered by both Professor Koenig[100] and Professor Thomas Kuhn.[101] This basic statement requires several qualifications. It is certainly true that Carnot mentions the impossibility of a perpetual motion machine, and takes this basic truth for granted. Again, as happened later in the case of Helmholtz, this in itself is not a sufficient condition for the formulation of a conservation principle, but it is a necessary one. It is also true that Carnot often speaks in a way that could be interpreted as if he meant that heat was motion: In a footnote we find, for example, the following:

The objection may be perhaps raised here, that perpetual motion, demonstrated to be impossible by mechanical action alone, may possibly not be so if the power either of heat or electricity be exerted; but is it possible to conceive the phenomena of heat and electricity as due to anything else than some kind of motion of the body, and as such should they not be subjected to the general laws of the mechanics? Do we not know besides *a posteriori*, that all the attempts made to produce perpetual motion by any means whatever have been fruitless?[102]

This passage is revealing. (Incidentally, La Mer quotes it too, without even mentioning that it is a footnote and not in the body of the text.) Let us examine the facts: it is a methodological comment, not a scientific one. Carnot surely read the Lavoisier–Laplace joint enterprise, and it is very probable that he was influenced by its methodological aspect. It is an ever-recurring pattern that scientists pay lip-service to a methodology which partially or even completely belies their own conceptual framework. In this case I should like to advance the view that Carnot was not an ardent believer in the material theory of heat even at this time, and when pressed by a methodological problem he was very much aware of it; but at this stage of his work he thought in caloric terms. This becomes clear whenever he passes to science proper, and we shall quote another short passage, which sounds even more caloric-minded than the one quoted so far. I did not include this quotation when considering 'strong evidence', since it is so typically unintentional:

The caloric developed in the furnace by the effect of combustion traverses the walls of the boiler, produces steam, and in some way incorporates itself with it.

If we try to unite these two approaches we will have to say simply that Carnot was in reality undecided. In the original 'Memoir' he says:

For the rest we may say in passing, the main principles on which the theory of heat rests require the most careful examination. Many experimental facts appear almost inexplicable in the present state of this theory.

The philosophical status of this undecidedness is very different from a non-committal attitude on methodological grounds. It is not that Carnot does not believe that there is a 'real' solution to the problem, and that he considers theories operationally—he simply does not know; and this is a great difference indeed! Before going on to the posthumous 'Notes' of Sadi Carnot, for the sake of fairness and objectivity we must at least mention why Callendar, La Mer and others tried to interpret Carnot's 'Calorique' as entropy. The reason is plain, since then we get a new theory which is true and modern, and we can then claim that Carnot was a modern physicist incidentally born one hundred years too early. This study deals neither with Carnot nor with the problems of entropy; consequently I shall avoid semantic details. Koenig and Kuhn have shown that this interpretation is false, and furthermore 'entropy' cannot systematically be substituted for 'calorique' for very often we would get complete nonsense. A similar argument will be presented concerning Helmholtz's 'Kraft', and there we will go into detail in order to prove the fallacy of the substitution approach. Let me remark only that the above approach seems undesirable and unjustified in the broadest historical terms. It is not the task of the historian of science to discover premature geniuses; furthermore even if such an interpretation would have been true, appreciation would not be greater in the eyes of every historian of philosophical concepts. This appreciation could be higher; finally, I do not believe that anyone is born a century before his time, or a century *after* his time for that matter. To end this digression with a commonplace (which is nevertheless true): every great mind is as much the creator of the century after him as he is a part and product of his own.

Carnot: the 'posthumous remarks'

Carnot died very young; his brother H. Carnot submitted a 'Lettre' to the Académie des Sciences. In this letter he states that the accompanying unedited fragments of his brother 's'ils n'apportent point à la Science des resultats nouveaux, témoignent que Sadi Carnot avait prévu avec une assez grande netteté que l'on a plus tard tirées de ses idées'.[103] This is probably the origin of the later myth, that Carnot had from the beginning been in full possession of the foundations of modern thermodynamics. Whether this was implied by his brother or not, I do not know, but it is certainly false. However, these notes do contain serious changes in Carnot's conceptions; there is a clearly formulated switch to the motion theory of heat, and an almost as clear formulation of the conservation of energy principle. It is not certain when the notes were actually written. In Mendoza's selection they are arranged according to the dating of Raveau; but since Carnot died in 1831 at the age of thirty-six, it is not very important to try to settle this question here. Had this been a study in priorities, this point would have necessitated a profound analysis. But it is enough to remind ourselves that these notes lay unnoticed till 1878. It is certainly true that having arrived at these conclusions later, it is more understandable that La Mer and others tried to read these thoughts back to the original 'Mémoire'. I certainly agree with Kuhn that 'the notes reflect on the brilliance of the memoir's author, not on the memoir itself. Since they refer to problems developed in the memoir, and are occasionally in fundamental conflict with it, most, or all of the notes must have been written after 1824.'

From these notes it becomes clear that Carnot knew very well the work of Rumford and Davy, and moreover (and in this he is nearer to the position of Helmholtz twenty years later) he knew the work that had been done in mechanics. He knew the conservation law of *vis viva* and he certainly understood the generality of the mathematical treatment involved. Had he been absolutely sure, even now, in the motion-theory of heat, and had he lived to perform the experiments he suggested for testing the heat-work equivalence, he might have discovered thermodynamics.

Up to the present time the changes caused in the temperature of bodies by motion have been very little studied . . . Is heat a result of a vibratory motion of molecules? If this is so, a quantity of heat must be unchangeable. . . . Can examples be found of the production of motive power without actual consumption of heat? It seems that we may find production of heat with consumption of motive power (re-entry of air into a vacuum for example). . . . What is radiant caloric? . . . Could a motion (that of radiant heat) produce matter (caloric)? Undoubtedly no; it can only produce motion. Heat is then the result of a motion.[104]

There is no need to comment on the breadth and originality of these notes. They speak for themselves. Carnot even deals with the one problem which Planck pointed out as the missing element in most of the early theories of conservation, namely with the idea that nothing can be lost without gaining something and not only the reverse of it:

If, as mechanics seem to prove, there cannot be any real creation of motive power, then there cannot be any destruction of this power either—for otherwise all the motive power of the universe would end by being destroyed—hence there cannot be any real collision between bodies.[105]

One of the most important of these notes is the one where Carnot suggests a repetition of Rumford's cannon experiment (actually he speaks of metal-drilling) 'but to measure the motive power consumed at the same time as the heat produced'. How is it possible that, saying so much, he does not comment on the fact that Rumford considered the heat generated by friction as inexhaustible? Is it possible that he too, by sheer hindsight, misinterpreted Rumford?

Much more could be said about Carnot. His work should be examined from the point of view of his exact use of language, the influence on him of his father, as far as terminology goes, and also the influence of Lazare Carnot's 'Fundamental Principles of Equilibrium and Movement', also to see whether the École Polytéchnique with its mathematical spirit, and its great teachers had a decisive influence on his thought! What part did Laplace play in S. Carnot's mental development? Why is it, that though he was not really committed to any model of heat, the strong positivistic influence of the École Polytéchnique seems to have passed him by? If I venture a guess not founded on evidence as far as Sadi Carnot is concerned —but in complete agreement with the general approach developed here— I would say that positivism did not have much to offer to a concept-creating scientist whose major contribution was not in the field of mathematical synthesis.

In the 1840s the conservation of energy was established by several of the great scientists of the time. Helmholtz formulated his general treatment in 1847. In 1850 Clausius enunciated the 'Clausius formulation' of the Second Law and the First Law as joint premises, and recognized their independent existence. One year later Thomson deduced Carnot's theorem from his own formulation of the First Law and showed that under the First Law the Clausius and Kelvin statements were equivalent.

NOTES

1. C. Møller, 'The Concept of Mass and Energy in the General Theory of Relativity', *DKNVS Forhandlinger* **31** (1958).

2. A. Katchalsky and P. F. Curran, *Nonequilibrium Thermodynamics in Biophysics* (Harvard, 1965), pp. 10–19.

3. Lord Kelvin, 'On a Universal Tendency in Nature to the Dissipation of Mechanical Energy'. *Trans. Roy. Soc. Edinburgh* **20** (1853), 261.

4. This is a paraphrase of the original Clausius formulation as given by Professor L. Tisza in 'The Evolution of Concepts of Thermodynamics' (ch. 1), *Generalized Thermodynamics* (M.I.T. Press, 1966).

5. T. S. Kuhn, 'Energy Conservation as an Example of Simultaneous Discovery', in M. Clagett (Ed.), *Critical Problems in the History of Science* (Wisconsin, 1955).

6. See La Mer, 'Some Current Misinterpretations'.

7. *Arch. Intern. d'Histoire des Sciences* **27** (1948), p. 630.

8. Professor Sanborn C. Brown has published several articles and two books on Count Rumford: (1) 'Count Rumford's Concept of Heat', *Amer. J. Phys.* **20** (1952), 331; (2) 'The Caloric Theory of Heat', *Amer. J. Phys.* **18** (1950); (3) *Count Rumford, Physicist Extraordinary* (Anchor Books, 1952); (4) *Benjamin Thompson, Count Rumford* (Pergamon Press, 1966).

9. See Koenig, 'History of Science'.

10. Hiebert, *Historical Roots of the Conservation of Energy*.

11. Francis Bacon, *Novum Organum* in vol. iv of Spedding's translation of *Bacon's Works*.

12. It also illustrates that Bacon did propose physical theories which had not been arrived at inductively out of the patient, endless, pointless experiments which can be performed even by the mediocre physicist—a picture which is often painted of Bacon's philosophy of science and which is not more than a mere caricature.

13. L. D. Patterson, 'Robert Hooke and the Conservation of Energy'. *Isis* **38** (1948), 151.

14. J. Tyndall, *Heat as a Mode of Action* (London, 1804), p. 39.

15. A. Lavoisier and P. S. Laplace, '*Mémoire sur La Chaleur*' (1780), éd. Gauthier-Villars.

16. ibid.

17. ibid.

18. See Lilley, 'Attitudes to the Nature of Heat'.

19. Robert Hare, *Compendium of the Course of Chemical Instruction* (1828), p. 5.

20. 'Evolution of Concepts' in *Generalized Thermodynamics*, p. 105.

21. I do not intend to go into details as to the work of Joseph Black and Lavoisier, but refer the reader to articles and books on the subject: (1) G. R. Partington, 'J. Black's Lectures on the Elements of Chemistry', *Chymia* **5** (1959), 130; (2) H. Guerlac, 'J. Black and Fixed Air, Etc.', *Isis* **48** (1957), 124, 433; (3) H. Guerlac, *Lavoisier* (Cornell, 1961); (4) D. McKie, *Lavoisier* (London, 1952).

22. R. A. Cleghorn, *Disputatio Physica Inauguratis, Theoriam Ignis Competentur* (Edinburgh, 1773).

23. See Brown, Caloric Theory'.

24. P. L. Dulong et A. T. Petit, *Ann. Chem. Phys.* **7** (1818).

25. See Brown, 'Caloric Theory'.

26. Count Rumford, *Mémoire sur la Chaleur* (Paris, 1804).

27. ibid.

28. S. C. Brown (ed.), *Collected Works of Count Rumford* (Harvard, 1968), vol. I, p. 23.

29. I. B. Cohen, *Franklin and Newton* (Philadelphia, 1956), pp. 222–34.

30. See *Benjamin Thompson, Count Rumford* (1966).

31. Brown (ed.), *Collected Works*.

32. ibid.

33. Planck, *Das Prinzip der Erhaltung der Energie* (1913).

34. See note 8 above.

35. *Benjamin Thompson, Count Rumford*, p. 15.

36. ibid. Taken from a collection of copies of letters from Rumford to Pictet in the American Academy of Arts and Sciences in Boston.

37. Masao Watanabe, 'Count Rumford's First Exposition of the Dynamical Aspect of Heat', *Isis* **50** (1959), 141.

38. Tyndall, *Heat as a Mode of Action*, p. 25.

39. There is no need to enter here into biographical detail. The most often quoted (1831) biography of Davy is by Dr Paris—John Ayrton Paris, *The Life of Sir Humphry Davy, Bart., Ll.D.* (2 vols., London, 1831).

Dr John Davy, Sir Humphry's brother has left three biographical works: (1) *Memoirs of the Life of Sir Humphry Davy, Bart.* (2 vols., London, 1836); (2) *The Collected Works of Sir Humphry Davy* (9 vols., London, 1839–40); vol. i is the biography; (3) *Fragmentary Remains, Literary and Scientific of Sir Humphry Davy, Bart., With a Sketch of His Life and Selections from His Correspondence* (London, 1858).

40. Davy, *Collected Works*, vol. i, p. 17.

41. ibid.

42. ibid., vol. ii, p. 28.

43. L. Pearce Williams, *Michael Faraday* (Basic Books, 1966).

44. J. T. Merz, *History of European Thought*, vol. 2, p. 103.

45. A. Lavoisier, *Traité Élémentaire de Chimie* (vol. i, Paris, 1789), p. 6.

46. A remark by Humphry Davy published in Nicholson's *Journal* shortly after the publication of the first essays, and quoted by John Davy (*Collected Works*, vol. i, (1935), p. 24.

47. ibid., vol. ii, p. 18.

48. The result of the experiment is the same if wax, tallow, resin, or any substance fusible at a low temperature be used; even iron may be fused by collision, as is evident from the first experiment.

49. da Andrade, 'Two historical notes', *Nature* (1935).

50. Davy, *Collected Works*, vol. ii, p. 117.

51. Davy, *Fragmentary Remains*, p. 52.

52. J. B. Stallo, *The Concepts and Theories of Modern Physics*, ed. P. W. Bridgman (Harvard, 1966), p. 109.

53. Davy, *Fragmentary Remains*.

54. All these quotations are from J. Davy's *Collected Works*, vol. iv, pp. 65–9.

55. Davy's Presidential Address to the Royal Society in 1820. Reprinted in the *Fragmentary Remains*, p. 231.

56. A. Wood, *Thomas Young*, Ed. Frank Oldham (Oxford, 1954), p. 319.

57. Davy, *Collected Works*, vol. i, p. 116.

58. ibid.

59. ibid.

60. ibid.

61. ibid.

62. ibid.

63. ibid., p. 16.

64. ibid., vol. i, p. 24.

65. ibid., vol. viii, p. 284.

66. ibid.

67. ibid., vol. viii, p. 286.

68. ibid.

69. ibid.

70. ibid.

71. ibid., vol. v.

72. ibid., vol. viii, p. 290.

73. ibid., vol. i, p. 155.

74. ibid., vol. viii, p. 337.

75. ibid., vol. i, p. 15.

76. Here, as in the case of Helmholtz's use of 'Kraft', we find in Agassi's 'Towards a Historiography of Science' a clearly stated awareness of the problem.

Further articles and books on Davy are: (1) Robert Siegfried, 'The Chemical Philosophy of Humphrey Davy', *Chymia* 5 (1959), 193; (2) J. Z. Fullmer, 'On the Poetry of Sir Humphrey Davy', *Chymia* 4 (1958), 2d; (3) The chapter in Pearce Williams' book (see (143)); and (4) A recent book, Sir Harold Hartley, *Humphry Davy* (London, 1966).

77. J. Z. Fullmer, 'On the Poetry of Sir Humphry Davy', *Chymia* 4 (1958).

78. ibid.

79. ibid.

80. ibid.

81. Davy, *Collected Works*, vol. i, p. 153.

82. ibid.
83. ibid., vol. viii, p. 340.
84. ibid.
85. ibid.
86. ibid., vol. i, p. 135.
87. ibid., vol. vi, p. 257.
88. ibid.
89. ibid., vol. i, p. 23.
90. ibid., vol. viii, p. 272.
91. L. Pearce Williams, *Michael Faraday* (Basic Books, 1966), pp. 60–73.
92. K. Coburn (Ed.), *The Philosophical Lectures of S. T. Coleridge* (The Pilot Press, London, 1949).
93. ibid., p. 399.
94. Davy, *Collected Works*, vol. i, p. 450.
95. Davy, *Fragmentary Remains*.
96. In *Essays in Honour of Herbert Evans* (California, 1943).
97. E. Mendoza (Ed.), *Reflections on the Motive Power of Fire by S. Carnot and other Papers* (Dover, 1960).
98. J. N. Bronsted, *Energetics* (New York, 1952).
99. 'Some Current Misinterpretations of N. L. Sadi Carnot's Memoir and Cycle III', *Amer. J. Phys.* **23** (1955).
100. 'History of Science.'
101. 'Carnot's Version of "Carnot's Cycle" ', *Amer. J. Phys.* **22** (1954), p. 91.
102. Mendoza, 'Reflections by S. Carnot', p. 12.
103. H. Carnot, 'Lettre Addressée à l'Académie des Sciences' (Paris, 1878).
104. S. Carnot. *Posthumous Manuscripts*, in Mendoza, 'Reflections', pp. 60–9.
105. ibid.

IV

PHYSIOLOGICAL
BACKGROUND

Helmholtz's physiological background

By education Helmholtz was a physician, and from 1839 he spent several years in Johannes Müller's famous laboratory, where the most brilliant thinkers in physiology and medicine worked together at the time:[1] Emile Du Bois-Reymond, Brucke, Virchow and the somewhat older Ludwig. Most of them were thorough reductionists, and under Müller's guidance they tried to solve the problem of the sources of animal heat; the problem itself had been formulated in Liebig's very much 'vitalistic' language.[2] What I wish to emphasize here is that they tried to reduce to mechanics a problem which belonged to biology, and was formulated with the help of concepts like 'vital force' or 'forces of life'. When coming to assess the influence of this background on Helmholtz's 'On the Conservation of Force' we have to bear in mind that this was his first paper in physics: his previous papers had been connected with the above-mentioned question of animal heat, and had been published two years and one year respectively before the 'Erhaltung'. No doubt, in those years he was preoccupied by the balance of 'forces' in the basic life processes. His papers on this topic were 'Über den Stoffverbrauch bei der Muskelaktion', 'Über das Wesen der Fäulniss und Gährung' and a review article for the *Fortschritte der Physik im Jahre 1845* called 'Bericht über die Theorie der PhysiologischenWärmeerscheinungen für 1845'.[3] It is interesting that the last one was written for a periodical in physics, and that Helmholtz himself when editing his collected scientific papers in 1881 classified it among the papers under the title: 'Zur Lehre von der Energie'. In 1847, the year of the publication of the 'Erhaltung', he published another article 'Über die Wärmeentwicklung bei der Muskelaktion'. Especially interesting is the article Helmholtz wrote in 1845 for the *Encyklopaedische-Handwörterbuch der medizinischen Wissenschaften*, under the item 'Wärme, physiologisch'. It reads like an expanded version of all the above-mentioned premises, for which the final conclusion is not drawn. He explains at length both the caloric theory of heat and the mechanical theory, and considers the latter as the correct one, but emphasizes its incompleteness.

He states very explicitly his reductionist views of physics to mechanics and also of life processes to physics. It is clearly stated that there must be a unique connection between the mechanical conservation law and the 'Life Forces'; only the belief in a basic conservation law in Nature is not stated in so many words, but can be read between the lines.

In later years Helmholtz repeatedly said that his 'On the Conservation of Force' was primarily written for the needs of physiology. What all this amounts to is, that though in his early youth he had read the works of Newton, d'Alembert, Lagrange, and Euler, he was then active in physiological research—his terminology is that of biology, and he formulated his problems in that language. It was thus natural for Ruecker to say about Helmholtz in 1894:

The study of medicine led him to the problem of the nature of 'vital forces'. He convinced himself that if as Stahl had suggested, an animal had the power now of restraining and now of liberating the activity of mechanical forces, it would be endowed with the power of perpetual motion. This led to the question whether perpetual motion was consistent with what was known of natural agencies. The Essay on the Conservation of Force was, according to von Helmholtz himself, intended to be a critical investigation and arrangements of facts which bear on this point for the benefit of the physiologist.[4]

Helmholtz's own testimony on this (very much tainted by his later inductivist preaching) and also the assessment of the influence of Johannes Müller on him can be found in the Ostwald biography.[5] The controversy in which Helmholtz took such a prominent part was about Liebig's 'vitalistic' theory, namely, whether the forces which were responsible for the production of physiological heat, were 'vital forces' or well-known physico-chemical processes.

Eighteenth-century traditions in 'vitalism'

Before taking a closer look at what Helmholtz did in this field—we are interested in his work prior to his 'Erhaltung der Kraft' paper—we have, however, to see more clearly the background and problem situation in physiology: what were the questions asked in Johannes Müller's laboratory, and in what terms were the questions formulated?

In the eighteenth century the gaps between the chemist, the physiologist, and the medical man were very small indeed. If we want to have an idea on the state of affairs in any of these sciences, the best way is to look up any standard textbook of the 1750s which goes under the name of 'physiology', or sometimes under that of chemistry, like Boerhaave's famous treatise. Albert Haller's *Outline of Physiology* appeared in 1747 and was almost immediately translated into English and French. One of his major problems, as of any of the older physiologists, was the problem of the origin of animal heat. This topic is historically covered by June Good-field's *The Growth of Scientific Physiology*[6] and in Everett Mendelsohn's *Heat and Life: Development of the Theory of Animal Heat.* In Haller's book we find the following explanation:

[heat arises] from the alternate extension and contraction, relaxation, and com-pression of the pulmonary vessels, by which the solid parts of the attrition that is made during expiration, as it is more rapidly moved and ground together during expiration.[7]

The importance of this passage resides in the fact that, as today, the non-physical sciences chose an accepted model from physics, on grounds which have nothing to do with physics, and afterwards treated that model as the basic truth. We have seen to what extent it was questionable—even in the 1800s—whether heat was motion or particles, and that certainly in 1750 the scientific character of the motion-theory of heat was very debatable. But for historical and also scientific reasons (inside physiology) this model has been accepted and was the ruling mode of explanation in physiology for the next hundred years. When on philosophical grounds the problem of vital forces arose in the 1820s, the motion theory of heat had in the meantime gained much stronger acceptance on physical grounds. Thus the physiological setting with its strong tradition (and the medical profession is one of the most tradition-minded; medicine was the only field of science which had in all ages cultivated the writing of its own history) joined in the 1820s the physical solution of the same problem, and thus created the necessary conceptual background for reductionism. This does not imply that the physiologists did not experiment; on the contrary, Epstein's *Testbook of Thermodynamics*[8]—which made me aware of Haller's part in these developments—quotes a passage from the same *Outline of Physiology*, which is very interesting in this sense:

Nor is it any objection to this [the above described theory that animal heat originated in rubbing of the solid parts of the blood] that water cannot be made hot by any friction. Nor in reality is this assertion true; for water, violent winds, and motion, as well as milk, acquire some degree of warmth.

It is thus easy to understand that the work of Laplace and Lavoisier con-firming Priestley's work on the now accepted combustion theory of respiration, found loud echoes among the physiologists. This also explains why—when Lavoisier and Laplace discovered experimentally that this theory was not completely satisfactory (i.e. that the oxidation of carbon found in carbohydrates of the blood is not enough to explain exactly the amount of heat produced)—physicists saw their discovery as an almost stunning blow to the motion-theory, while physiologists looked for an explanation in the framework of that theory. Lavoisier tried to explain the discrepancy by suggesting that it was due to the oxidation of hydrogen, and he happily clung to the caloric theory, which, as we have seen in the previous chapter, was well founded enough. This discrepancy in the experiments of Lavoisier and Laplace, or as some called it 'Dr Crawford's theory', became the starting point of most physiologists who worked around the 1820s to 1840s.

Some approached the problem with the basic presupposition that phenomena of life are essentially different from those of the inorganic world; while others regarded the very fact that animal heat is caused by friction (inaccurate as the results were known to be) as committing them to the idea that there is a basic similarity between all material things on earth, be it live or dead matter. How this leads to the problems of vitalism needs no explanation, though we can hardly pause to describe the very complexity of this problem. I wish only to emphasize that the problem cannot be dealt with in oversimplified philosophical terms. For neither physics nor philosophy has provided an unambiguous answer, not then and not now. One would expect that the idealistic tradition in Germany, originating with Leibniz (and in the form of the basic beliefs of the Naturphilosophie), would support a mysterious vital principle which is one of the forces of nature. This is indeed true for many investigators; but what about those who on the same Leibnizian grounds rejected any basic difference between dead and living matter and wanted to see the same laws of nature applied to both, on the basis of their philosophical commitment to a great unity? Naturally, the solution adopted by many, some of them the greatest scientists of their age, was that all forces had a mysterious quality, force being something which is conserved in nature. The clarification of this situation will come, in my opinion, with a profound study of the Leibnizian conception of biology; for Leibniz's monads, those vague entities—which (as is mentioned in the introduction) were not meant to be translated into our terminology, and cannot be made understandable in those terms—were also a solution of Leibniz's physiological questions. They constituted the great unity between soul and matter; they were centres of his 'live' force, and he meant what he said. In other words, in biology Leibniz was a 'vitalist' and his solution was basically to consider all forces of nature as 'live forces', thus retaining the unity all-important to him.[9]

The confused traditions described in the preceding pages show the true reason why historians of science cannot make up their minds whether some of the greatest figures of nineteenth-century physiology and animal chemistry were vitalists or not. We see this of Berzelius, Liebig, Johannes Müller and Ludwig. The whole spectrum of interpretations can be applied to their views, from extreme vitalism to extreme reductionism. The fact that this basic philosophy did not interfere with what a scientist knew about the physics of organism is perfectly illustrated in Joseph Henry's *Remarks on Vitality*, written as an answer to the Rev. H. H. Higgins's *On Vitalism*. At the beginning the reader expects a thoroughly reductionist approach, when suddenly the vital forces make an appearance; it is worth quoting at length:

In the early study of mechanical and physiological phenomena, the energy which was exhibited by animals or in other words, their power to perform what is technically called work, that is to overcome the inertia and change the form of matter, was referred to the vital force. A more critical study of these phenomena has however shown that this energy results from the mechanical power stored away in the food and material which the body consumes; that the body is a machine for applying and modifying power, precisely similar to those machines invented by man for a similar purpose. Indeed, it has been shown by accurate experiments that the amount of energy developed in animal exertion is just in proportion to the material consumed. To give a more definite idea of this, we may state the general fact that matter may be considered under two aspects, namely matter in a condition of power, and matter in a state of entire inertness. Again, coal and other combustible bodies consist of matter in a condition of power, and in their running down into carbonic acid and water, during their combustion, evolve the energy exhibited in the operations of the steam engine. The combustible material may be considered the food of the steam engine, and experiments have been made to ascertain the relative economy in the expenditure of a definite amount of food in the natural machine and the artificial engine. The former has been found to waste less of the motive power that the latter.

In pursuing this train of investigation the question is asked, 'Whence does the coal or food derive its power?' The answer is, that these substances are derived from the air by the decomposing agency of the impulses from the sun, and that when burned in the engine or consumed in the body they are again resolved into air, giving out in this resolution an amount of energy equivalent to that received from the sun during the process of their growth.

In the gradual development of the principles we have given there has been a tendency to extend the views we have presented too far, and to refer all the phenomena of life to the mechanical or chemical forces of nature. Although it has been, as we think, conclusively proved that from food, and food alone, come all the different kinds of physical force which are manifest in animal life, yet as the author of the preceding paper has shown, there is something else necessary to life, and this something though it cannot properly be called a force, may be denominated the vital principle. Without the influence of this principle the undirected physical powers produce mechanical arrangements and assume a state of permanent equilibrium by bringing matter into crystalline forms or into a condition of simple aggregation, while under its mysterious influence the particles of matter are built up into an unstable condition in the form of organic molecules. While therefore we may refer the changes which are here produced, or in other words the work performed, to the expenditure of the physical powers of heat, chemical action, &c., we must admit the necessity of something beyond these which from the analogy with mental phenomena, we may denominate the directing principle. Although we cannot perhaps positively say in the present state of science that this directing principle will not manifest itself when all the necessary conditions are present, yet in the ordinary phenomena of life which are everywhere exhibited around us, organization is derived from vitality, and not vitality from organization. That the vital or directing principle is not a physical power which performs work, or that it cannot be classed with heat or chemical action is evident from the fact that it may be indefinitely extended—from a single acorn a whole forest of oaks may result.[10]

I chose this quotation for another reason—the date! Almost twenty years had passed since the clear establishment of the principle of conservation of energy—and Henry actually uses the term 'energy' correctly—but this did not interfere with his basic beliefs!

What was this 'vital force' in which they believed? Again, as we have seen in the case of the caloric, there were weighty reasons in favour of its acceptance. In an article 'Berzelius und die Lebenskraft', Bent Søren Jørgensen tries to reduce the problem to clear questions:

The problem of the existence of such a force could be solved by answering the following four questions, all posed in the chemical literature, but not in this context:

(1) Do the general laws of chemistry apply to chemical conversions in living nature?

(2) Do the general laws of chemistry apply to the products of chemical conversions in organic nature, if said products are being investigated outside the organism?

(3) Is it possible to imitate, *in vitro*, the products of the life processes (the actual synthetic problem)?

(4) Does the production of organs require a special vital force?[11]

Only the first question can be answered in the affirmative in view of what we know today. Question (2) is connected with the problem of interference of the experimenter, which is certainly open to criticism; the third one explains why the famous urea synthesis of Wöhler, in 1828, did not eliminate the problem from science. This problem has been extensively treated by Paul Walden[12] and by McKie.[13] About the fourth we have not much information; it puts us into the centre of the violent debate in embryology which was at its height in the 1930s and is still far from any solution. In any case, Berzelius's highly important experimental work in organic chemistry contributed to the opinion that in reality he was a reductionist. It is not worth quoting all the relevant sources here. Enough that Partington in his *History of Chemistry*[14] (the shorter one) and Pledge,[15] do not even mention this philosophical preoccupation. Partington goes even further—he formulates the description of Berzelius's contributions to science in such a way, that one could not imagine any other attitude but pure positivism in the twentieth-century style. As against this let us look at Berzelius's own words in his fascinating introduction to the organic part of his *Lehrbuch der Chemie*:

In living nature the elements appear to obey quite different laws than in the dead state. . . . Discovering the cause of this difference . . . would provide a key to the theory of organic chemistry. However, the latter is skewed in such a way, that at present, there is no hope for a solution. We must, nevertheless, strive to approach this cognition; for at some time we shall certainly reach our objective, or remain at a specific point beyond which human research cannot be extended. When regarded as the object of a chemical examination, a living body is a workshop in which numerous chemical processes take place, whose final result is the creation of phenomena, the totality of which we call life. . . . After a certain slow-down in the processes they finally cease, and from this moment the elements of the previously living body begin to obey the laws of unorganic nature.

Every organic body is thus distinguished from an unorganic one, in that the first has an observable beginning, that it develops, decreases, ceases to function and is destroyed. In contrast, the unorganic body was there before us, and consistently continues to exist in such a way, that its nature cannot be changed, regardless of circumstances.

The unorganic elements of the organic body can also not be destroyed, but the actual nature of that body is irrevocably destroyed. . . . Consequently, the nature of the living body does not originate in its unorganic elements, but in something else, which disposes the unorganic elements common to all living bodies, in order to create a result, which is specific and characteristic for each species or kind.

This something, which we call the life force, lies completely beyond the unorganic elements, and is not one of its original properties, such as weight, impenetrability, electrical polarity etc. . . . A force, incomprehensible to us and alien to dead nature has introduced this something into the unorganic mass . . . in admirable variety, and calculated for specific purposes with the highest wisdom. . . . All effects originate in what we call forces; the latter (like the will) strive to be implemented or satisfied, in order to attain a rest—after satisfaction—which cannot be disturbed. . . . We do not see how this very striving of unorganic matter towards an indifferent state of rest via the need to satisfy alternating forces, can maintain said matter in a state of incessant activity; . . .[16]

There is one aspect of Berzelius's scientific work which, I suspect, is in intimate connection with vitalism: his pioneering work on crystals, where he repeatedly drew attention to the concept of 'organization'.

Liebig

The same situation is true for the case of Liebig. Timothy O. Lipman deals with the problem of Liebig's vitalism in two articles—'Vitalism and Reductionism in Liebig's Physiological Thought' published in *Isis*, and another article which interests us more for our purposes, 'The Response to Liebig's Vitalism'.[17] He mentions that many of Liebig's contemporaries ignored this aspect of his rich scientific activity, even such people who were aware of the importance of the question like Theodore Bischoff or Berzelius. As we shall see later even Helmholtz did not condemn this vitalism unambiguously. Du Bois-Reymond, who is often quoted on this issue as a violent opponent of Liebig, expressed himself sharply only in later years, when it had become clear that Liebig's vitalism is in opposition to the programme of the extreme reductionists, like himself, Brücke, Ludwig or Helmholtz. Ludwig formulated the reductionist programme in the famous statement: 'We four imagined that we should constitute physiology on chemico-physical foundations, and give it equal scientific rank with physics' (1847). We shall come back soon to Johannes Müller, but let me mention in passing that Brucke, Du Bois-Reymond and Helmholtz were students of Müller, and they did not seem to have recognized how near Müller himself was to those abused vitalistic views. Once again the origins of a famous philosophy of science must carefully be explored since having turned out to be successful in the eyes of many, it is presupposed that it must have been reached as an experimental conclusion, based on years of experience. When Du Bois-Reymond began his scientific work, he was less than twenty years old. In that very year, in 1842, he made his famous statement: 'Brücke and I have sworn to make prevail the truth that in the organism no other forces are effective than the purely physical chemical . . .'. A most clearcut case of an *a priori* approach! All this has been explored in articles by Paul F. Cranefield.[18] I disregard here on purpose those extreme reactionaries who thought that chemistry or physics would never have anything to do with physiology or medicine. One of them was the American physiologist Charles Caldwell, who, in an outline of a lecture course to be given at the University of Pennsylvania in the 1820s, wrote:

No sooner has life forsaken organized matter than chemistry invades it—not before.[19]

When Liebig's work appeared in the 1840s Caldwell attacked it violently because it was not vitalistic enough for him. In his eyes Liebig seemed an extreme reductionist.

Of special interest is Liebig's concept of 'force' and I shall deal with it here shortly, but in principle the same way as with Helmholtz's 'Kraft' in the next chapter. Concerning the general ambiguity of the terms 'vital forces' and 'Kräfte' in physiology there is an illuminating discussion in June Goodfield's *The Growth of Scientific Physiology*. She quotes the following passage from Liebig:

In the animal ovum as well as in the seed of a plant, we recognize a certain remarkable force, a vital source of growth. . . . This force is called the *vital force, vis vitae or vitality*.[20]

Miss Goodfield continues:

No explanation is given here of force: the paragraph as it stands is the description of a potentiality. But if we read further, we find that he does mean more: he believes that there is a unique agency operating in living material different in nature and manifestation from any other. . . . His vital force like Bichat's vital properties, has no connection with the soul or mind, and is directly open to experimental study. . . . Nor is this vital force *immaterial*. . . . Liebig introduces the word 'force' into his physiological framework . . . treats it as a kind of central force, capable of giving rise either to resistance or to motion: 'The chemical force, which kept the elements together acted as a resistance, which was overcome by the active vital force. Several things must be noted about this passage: first of all two ambiguities. To begin with the word 'resistance' appears sometimes to mean 'mutual physical pushes and pulls' and sometimes the capacity to slow down or prevent the occurrence of 'physical or chemical processes'. Of course, for anyone thinking in terms of central forces, the 'invincible resistance of a compound' to the action of the 'decomposing agent' would seem explicable only by supposing the decomposing force to be weaker than the binding force. . . .[21]

Full justice is done here to the difficulty, and the conceptual muddle is very clearly explained. However, and this is the reason that I have quoted at such length, I cannot accept the treatment of the problem as being merely verbal:

Yet the very word 'Kraft' which Liebig used is profoundly ambiguous . . .: 'The change of matter, the manifestation of mechanical force, and the absorption of oxygen are in the animal body so closely connected with each other, that we may consider the amount of motion and the quantity of living tissue transformed, as proportional to the quantity of oxygen inspired and consumed in a given time by the animal. For a certain amount of motion, for a certain proportion of vital force consumed as mechanical force, an equivalent of chemical force is manifested.' Suppose that for the word 'force' with its precise twentieth-century associations we here substitute the word 'energy'. The passage at once becomes clear and acceptable. Liebig's word 'Kraft' in fact carried the meaning both of 'force' and of our 'energy'. In 1840, the distinction between the two had not yet been made clear and the word 'Energie' did not appear in scientific papers until some twenty years later.[22]

It is precisely this sort of substitution which cannot be performed consistently. Not only was the distinction not made between force and energy, but many sentences only have meaning if we leave 'Kraft' intact, exactly as in English we have to leave Faraday's concept of 'Force' in the framework of his own definition, otherwise his whole essay 'On the Conservation of Force' becomes meaningless. I shall show this in more detail in the case of Helmholtz, but let us look at another passage from Liebig:

This absorption of oxygen occurs only when the resistance which the vital force of living parts opposes to the chemical action of the oxygen is weaker than the chemical action.[23]

On the 'substitution theory' we would have to write here 'energy of life' or 'heat' or even leave it as 'vital energy'. But could any sort of energy (a well-defined scalar quantity) 'oppose a resistance' to another action, whatever the 'action' may mean: and could we meaningfully use the expression 'energy is weaker than'? Moreover, at the risk of seeming pedantic, it was hardly the unavailability of the term 'energy' which prevented its being used. In scientific literature Thomas Young introduced it with a clearly defined meaning—he called *'vis viva'* simple energy in his 1807 'Lectures on Natural Philosophy'. Thomson and Tait used it in various articles and in their famous textbook; Helmholtz used it sometimes in the same vague sense as 'Kraft':

... weil wir die Proteinverbindungen überall als Träger der höchsten Lebensergien finden ...
... since we find the protein compounds everywhere as the bearers of the supreme vital energies ...[24]

On the contrary, here, even more than in physical discourses, the ambiguity of the word 'Kraft' was the very condition for maintaining an intelligible discussion: the physiological problem of the sources of animal heat could have been resolved only after the physical law of the conservation of energy had been formulated, that is, only after the fixation of the concept of energy. Simultaneously as the physical balance of energies had been measured for living bodies, a thorough vitalist had to give up this battle as lost and to transfer the idea of a 'special vital something' from the problem of physiological heat to another, new, and as yet unsolved, question.

Having shown that the scientific language of Helmholtz's contemporaries was rich in ambiguities of this sort, we can find them in his papers too:

Eine der höchsten, das Wesen der Lebenskraft selbst unmittelbar betreffenden Fragen der Physiologie. Nämlich die, ob das Leben der organischen Körper die Wirkung sei einer eigenen, sich stets aus sich selbst erzeugenden, zweckmässig wirkenden Kraft, oder das Resultat der auch in der leblosen Natur thätigen Kräfte, nur eigenthümlich modifiziert durch die Art ihres Zusammenwirkens, hat in neuerer Zeit besonders klar in Liebig's Versuch die physikalischen Thatsachen aus den bekannten chemischen und physikalischen Gesetzen herzuleiten, eine viel concretere Form angenommen, nämlich die ob die Kraft und die in den Organismen erzeugte Wärme aus Stoffwechsel vollständig herzuleiten seien oder nicht.[25]

One of the supreme questions of physiology, directly affecting the very essence of 'Lebenskraft', namely whether the life of organic bodies is the action of a purposefully acting 'Kraft', which has the power to generate itself, or the effect of 'Kräfte' which are active also in lifeless nature, but peculiarly modified through their interactions has recently achieved a much more concrete form in Liebig's attempt to derive the physiological facts from known chemical and physical laws. This new form of the question is: can the 'Kraft' and heat generated in the organism be entirely derived from the metabolic process?[25]

It is not really necessary to demonstrate the word by word untranslatability of the above passage into modern terminology. There is, however, another point in this passage worth mentioning. We have stated above that Liebig is generally considered a 'vitalist'. Here it seems as if Helmholtz says the contrary; but if we turn to another passage, this time in his review of Liebig's work, we understand what is happening:

Liebig, in seiner *Thierchemic* . . . stellt die theoretische Forderung auf, dass der Ursprung der Wärme, als eines Princips, welches einem gewissen Kraftequivalent entspräche, nicht aus nichts, sondern nur anderen Kräften hergeleitet dürfte.[26]

Liebig, in his *Animal Chemistry*, postulates the theoretical demand, that the origin of heat, as a principle, which corresponds to a 'Kraftequivalent', can be derived only from other 'Kräfte' and not out of nothing.[26]

Liebig, who was talking of 'vital forces' which, as we have seen in the passages quoted from Goodfield's book, were neither immaterial nor to such an extent metaphysical as to have something in common with soul or mind, seems to Helmholtz to be working in line with his reductionist demands! I would rather overemphasize this point than let it go unnoticed. In 1845 Helmholtz's concepts were so vague and in a 'state of flux', that he could even accommodate in his conceptual framework some 'vital forces', and believe that as such they could be reduced to physico-chemical terms!

Johannes Müller

The other great centre of physiology was the laboratory of Johannes Müller. Müller's work was mainly on the physiology of the senses—a field which by its very character is deeply connected with philosophical attitudes; the connecting link is the theory of perception—itself a combination of science and philosophy—with the boundary very unclear. It is here that Dubois-Reymond, Brücke and Helmholtz met, and it was with Müller's guidance that the first neuro-physiological problems of Helmholtz were formulated. Müller's central interest was to work out ideas conceived by him when he was a very young man, what became known as Müller's 'law of the specific energies'. The law:

energies of the light, the dark, and the coloured are not immanent to the external objects, (i.e.), to the causes of the stimulation, but to the visual sense substance itself; that the visual sense substance cannot be affected without being active in terms of its innate energies as the light, the dark, or the coloured; that light, shade, and colour do not exist for the senses as a finished and external something, so that the sense when struck has only the sensation of it, but that the visual sense substance itself if activated by any stimulus, whatever its kind, brings its affection to sensation in terms of the energies of the light, the dark, and the coloured.[27]

His theories were incorporated in two great works: *Comparative Physiology of the Visual Sense of Man and Animals* and the *Textbook of Physiology*. Müller concluded that our sensations do not present a true impression of the external world; but the important place where our impressions are formed was for him the sense-organ—in the visual case the eye. Helmholtz was, in a way, the follower of Müller, but he tended to give more and more importance to the central nerve organ, the brain. Also, for reasons of clarity, Helmholtz changed the Müllerian terminology from 'energies' to 'signs'. We cannot penetrate deeply into the philosophical background of Müller's doctrine, and it will suffice to quote one more passage from the above-mentioned article:

Müller's biographer, Haberling, has revealed him as an originally contemplative, speculative, and metaphysical, if not mystical, mind—shaken and disturbed by the intensive auto-observations and auto-experiments of his early years, who, suffering a profound depression in 1840, finally turned away from his original subjective method of exploring human nature in order to reach what he considered to be the firmer ground of comparative anatomy. But an interpretation of the law of the specific energies of the senses in terms of a regional activity of a given structure would be incompatible with those principles of morphological investigation which Müller himself adopted and outlined in his inaugural academic lecture as early as 1824, i.e. two years prior to the publication of his *Comparative Physiology of the Visual Sense of Man and Animals* and of his first version of the law of the specific energies of the senses. It remains true, he said in his inaugural lecture, that the anatomist studies details, while the more philosophically oriented morphologist refuses to consider living forms as isolated and finished units. He warned against the danger of accumulating a chaos of anatomical knowledge lacking living thought. Nature should be captured in her living state of becoming and generating. He referred to Caspar Friedrich Wolff and his *vis essentialis*, to Kielmeyer and, above all, to Goethe and his doctrine of metamorphosis, according to which each finished part is the carrier and representative of an original design or archtype.

Subjective experiences emerge from the law of the specific energies of the senses as the ultimate source and frame of reference of man's vision, for which the external *objects* provide only the 'conditions'. According to Müller, man, reaching out for objective reality, cannot encompass more than his own sensations. 'We do not know the essence of the outer objects and that which we call outer light; we only know the essences of our senses; and of outer objects we only know of their action on us in terms of our energies.' We identify in this restriction of our experimental world to its subjective aspect of Kantian heritage which Müller never denied and even openly acknowledged. It remains true, however, that the term and concept 'subjective', when it assumes its authentic meaning as it is embodied in the *Critique of Pure Reason*, encompasses much more than just man's sensations; in fact, it encompasses space and time, causes and effects, and all of the pure concepts of understanding.[28]

Müller still awaits a profound study; my only conclusion here is, that his influence on Helmholtz was very great and that he himself was a philosophically thinking man, undoubtedly much influenced by Kant, Goethe and many others. (This is not further developed here, but in addition to the above source, the reader can turn to Ostwald's biography,[29] or even to the few remarks, however biased they might be, by Driesch, in his *History of Vitalism*.)

Back to Helmholtz

Helmholtz's own contribution to physiology mirrors well all these developments. His first serious work was his dissertation, which, on Müller's demand actually took longer than intended by Helmholtz. This was Helmholtz's discovery that nerve fibres originate in the ganglia. The task had been proposed by Müller, and was in line with Müller's investigations on the anatomy and physiology of the nervous system. The thesis appeared in Latin: *De Fabrica Systematis Nervosi Evertebratorum*, 1842. At this time Helmholtz was influenced very much by his master, and it is interesting to see how his biographer, Leo Koenigsberger, evaluates this period in his life, his master and work. It is noteworthy what a confused view the otherwise very clear Koenigsberger has of the metaphysical foundations of Müller's physiology:

Müller had indeed emancipated himself from his earlier and essentially metaphysical views of the nature of life, and demanded an empirical foundation for all scientific concepts, but, as his pupils recognized, he had been unable to free himself entirely from the traditions of nature philosophy and from metaphysical conceptions. Under his influence Helmholtz endeavoured to lay the foundations of a strict physical science by ascertaining the facts in certain definite problems, thereby co-operating with the ceaseless efforts of his master.[30]

He had emancipated himself, but he was unable to free himself entirely! This attitude was also that of Helmholtz. He never attacked Müller's 'vitalism' and I believe that he never saw it in this light. We have seen at the beginning of this chapter that Helmholtz was also ready to accept Liebig's 'vitalism' for the time being. This, I do not interpret as simple blindness or prejudice, but rather a kind of bewilderment, especially because both from conversations and from Müller's writings, Helmholtz must have detected the Kantian influence on Müller, which at this time he even consciously accepted. At the same time, Helmholtz was already occupied with other physiological problems, with the question of the origins of animal heat, and at the background of his mind, there was already a metaphysical commitment to some sort of conservation law. The positivistic Koenigsberger, who does all in his might to support Helmholtz's claim to the fact that he was a 'hardened' empiricist—and

that all his important conclusions must have been drawn inductively—
gives himself away in an original fashion. (It would be interesting to know
whether Helmholtz ever talked with him on this subject in these terms.)

But in the mind of Helmholtz, the conflict between realistic and metaphysical
principles had become a resolute fight against the dominating ideas in a wider field
than physiology: a vanishing vital force for which nothing was substituted appeared
to him a physical paradox—disappearance of energy and matter was unthinkable.[31]

Helmholtz was twenty-two at the time, and this was his second scientific
undertaking, the first done without a supervisor. He had agreed to a 'vital
force', he certainly knew that this was a vague, mathematically unformulated concept, and he was convinced of its conservation. The last sentence
betrays Koenigsberger's motive—he sees in this Helmholtz's clear vision
of the energy concept and the principle of its conservation! But what had
this Liebig-style 'vital force' to do with the physical principle as it became
established in Helmholtz's work two years later? If I doubt Koenigsberger's objectivity in one direction, let me not overrely on him in a way
which would prove my point. There is one more proof—direct, and thus
unlike most of the argument in this study—in which Helmholtz himself is the speaker. Klein had accompanied Helmholtz on his American trip
in 1893, and wrote a letter to Koenigsberger describing this trip, at the
latter's request. There we find:

Auf der Rückreise [schreibt mir Klein] hatte der Verkehr ungezwungenere Forme
angenommen. Ich zeigte Helmholtz die Correcturbogen meines 'Evanstor
Colloquium', über die er dann allerdings nur sagte, er verstehe im Allgemeinen,
was ich beabsichtige. Ein anderes Mal sprach ich mit ihm über die Nothwendig-
keit einer technischen Physik im Sinne der bald hernach von mir bethätigten,
Ihnen bekannten Auffassung; er schien diese Auffassung nicht eigentlich zu theilen,
sondern die bestehende Ausbildung an der Universität für ausreichend zu halten.
Ein anderes Mal wieder kam die Unterhaltung auf die Geltung des Princips der
kleinsten Wirkung in der Physik, wobei ich dieselbe mit derjenigen des Gesetzes
der Erhaltung der Kraft parallelisirte; Helmholtz erwiderte, beide Dinge seien für
ihn in der That insofern ganz analog, als sie ihm bei Abfassung seiner bezüglichen
Schriften beide von vornherein ganz selbstverständlich vorgekommen seien. . . .[32]

This is taken from the German edition of the Koenigsberger biography;
the passage is missing in the shortened English version.[33]

   The 'conflict between realistic and metaphysical principles' in the mind
of Helmholtz (in the above-quoted passage by Koenigsberger) relates to
the assumption that Johannes Müller

did not attempt to disguise the inconsistency of his position, and as a result the four
gifted young investigators, Brücke, Dubois-Reymond, Helmholtz and Virchow
were all striving to develop a logical and unified physiology according to the
principles of exact investigations.

This formulation, in view of the above argument, seems to me very doubtful.

Let me refer once more to Helmholtz in another paper. We saw that his 'On the Nature of Fermentation and Putrefaction' was meant primarily to support Liebig on a form of vitalism—at this time Helmholtz accepted the Liebig style of vitalism, namely, the one which tried to reduce most of the biological phenomena to physico-chemical processes, while retaining the function of an unspecified 'vital force'. It is this double meaning of the 'Lebenskraft' which makes the debate on Liebig's philosophy so lively and interesting. Here Helmholtz found that the transformations known as fermentation and putrefaction are not the result of chemical action, as supposed by Liebig in his previous studies. He showed that putrefaction can occur independently of life, but that

it offers a fertile soil for the development of nutrition of living germs, and is modified in its aspects by them.

He drew the parallel clearly between fermentation and the processes of life, and thus became a clear forerunner of Pasteur. But for us the interesting result is, that though on the one hand Helmholtz did not reject vitalism completely, yet on the other, his results seemed to point even stronger in the direction of vital forces than Liebig's own assumptions. However, his results certainly did not clash with any conservation ideas; which merely proves that we cannot simply say that the metaphysicians were vitalists, while the 'good' scientists, who were positivistic and experimentalists, were necessarily reductionists. But I have been grinding this axe too long.

His last study before the publication of the 'Erhaltung der Kraft' was very near that topic. He wrote for *Fortschritte der Physik* the study mentioned at the beginning of the chapter, consequently reaching the brink leading to the establishment of the conservation of energy principle.

NOTES

1. E. Mendelsohn, 'The Biological Sciences in the Nineteenth Century: Some Problems and Sources', *History of Science* 3 (1964), p. 36. Here an interesting distinction is made between the German and French reductionists, respectively represented by Müller and Claude Bernard.

2. It remains conclusively to be decided whether J. Müller and Liebig were really 'pure reductionists' or 'pure vitalists' and how they developed in this respect.

3. All three papers in H. v. Helmholtz's *Wissenschaftliche Abhandlungen* (2 vols., Leipzig, Johann Ambrosius Barth, 1882). Another rich source for biographical material and for the study of change in Helmholtz's philosophy of Nature is *Vorträge und Reden* (2 vols., Braunschweig, 1903)

4. A. W. Ruecker, F.R.S., 'Helmholtz', *Ann. Rep. Smiths. Inst.* (1894), p. 709.

5. W. Ostwald, *Grosse Männer* (Leipzig, 1903).

6. J. G. Goodfield, *The Growth of Scientific Physiology* (London, 1960). E. Mendelsohn, *Heat and Life: Development of the Theory of Animal Heat* (Harvard, 1964).

7. A. von Haller, *Outline of Physiology* (1747).

8. S. T. Epstein, *Textbook of Thermodynamics* (New York, 1934).

9. On another level, one would expect that those physicists who succeeded in explaining so many of the biological phenomena in physico-chemical terms would all belong to an extreme reductionist school; but this is not the case. Not even today—nobody ever dreamt that so many of the previously hidden phenomena of life would be revealed by the work of the modern molecular biologists. And yet, among those not all are what we would have called 'reductionists' in nineteenth-century terms. Some of the very greatest believe that some of the phenomena of life will never be explained in physico-chemical terms, and that the world of physics and chemistry only advances the boundary. Others, nearer in temperament to what the nineteenth century would have called 'vitalists', shift the problem, which in their opinion will never be reduced to physics. It is not relevant that no working scientist today would agree to be called a 'vitalist' without taking offence; terminology has changed, and the scientific community pays even more lip-service to a shared philosophy of science than ever before. Moreover, these problems are nowadays couched in a jargon of technical philosophy. Just to mention some of them—whoever talks of 'biological complexity' as against 'physical simplicity', has, in my opinion, reformulated the nineteenth-century dialogue on vitalism. The question always boils down to the problem: is there any qualitative difference between live phenomena and those of dead matter? Some will always answer in the affirmative, while others will not.

10. *Ann. Rep. Smiths. Inst.* (1866), p. 386.

11. B. Søren Jørgensen, 'Berzelius und die Lebenskraft', *Centaurus* 10 (1964), p. 258.

12. Paul Walden, *Naturwissenschaften* 16 (1928).

13. D. McKie, *Nature* 153 (1944).

14. F. R. Partington, *A Short History of Chemistry* (London, 1957).

15. H. T. Pledge, *Science since 1500* (Harper, 1959).

16. J. J. Berzelius, *Lehrbuch der Chemie* (Wohler, 4 vols., Dresden, 1825–31).

17. T. O. Lipman, 'The Response to Liebig's Vitalism', *Bull. Hist. Med.* xl (1966), p. 511.

18. Paul F. Cranefield: 'The Organic Physics of 1847 and Biophysics Today', *J. Hist. Med.* xii (1957), p. 407; 'The Philosophical and Cultural Interests of the Biophysical Movement of 1847', *J. Hist. Med.* xxi (1966).

19. H. S. Klicstein, 'C. Caldwell and the Controversy in America over Liebig's 'Animal Chemistry', *Chymia*.

20. Goodfield, *Growth of Scientific Physiology*, pp. 135–45.

21. ibid.

22. ibid.

23. ibid.

24. H. v. Helmholtz, 'Über den Stoffverbrauch in der Muskelaktion' in J. Müller's *Archiv* (1845), p. 82; or *Wissenschaftliche Abhandlungen* (vol. 2), p. 745.

25. ibid.

26. H. v. Helmholtz, 'Bericht über die Theorie der Physiologischen Wärmeerscheinungen für 1845', *Fortschritte der Physik für 1845* (appeared 1847), p. 347; or in *Wissenschaftliche Abhandlungen* (vol. 1), p. 1.

27. This translation is taken from Walter Riese and George E. Arrington, 'The

History of Johannes Müller's Doctrine of the Specific Energies of the Senses: Original and later version', *Bull. Hist. Med.* 37 (1964).

28. ibid.

29. Ostwald, *Grosse Männer.*

30. Leo Koenigsberger, 'The Investigations of Hermann von Helmholtz on the Fundamental Principles of Mathematics and Mechanics', *Ann. Rep. Smiths, Inst.* (1890).

31. ibid., p. 25.

32. Koenigsberger, *H. von Helmholtz*, vol. iii, p. 90.

33. As to this, see Robert S. Cohen's review of the English edition of Koenigsberger's book; to be published.

V

'DIE ERHALTUNG
DER KRAFT'

The 'simultaneous discoverers'

This is not a study in priorities. It is well known that several great scientists in the 1840s arrived at the principle with better or worse formulations, but in such a way that none of them was accepted immediately as the originator. It was pointed out in the introduction, that the principle was first described, in all its generality in mathematical terms by Hermann von Helmholtz in his 1847 paper 'Über die Erhaltung der Kraft'. This made the principle finally acceptable to the scientific world, but only after this acceptance, did people begin to discover it in the work of others. Our study is meant to demonstrate why only Helmholtz arrived at the desired clarity, what the necessary ingredients were, why all these factors converged in Germany, and finally the influence of his personality. This order of things is purely methodological, and does not imply that I attribute greater importance to social-external factors than to the personality. On the contrary, personality is a sufficient condition; but to have become such a personality implies all the other factors, apart from genetics, which were considered here in turn as necessary conditions. We have seen the double tradition in (vectorial and scalar) mechanics which bothered Helmholtz, the state of scientific affairs in the theory of heat, the physiological background, and the fact that Helmholtz was enough of a mathematician to have mastered the language in which the principle had to be expressed. The problem of simultaneous discovery has been dealt with in detail by Professor Thomas Kuhn.[1] I shall not refer to this paper in detail, as most of the points upon which we disagree have already been covered. Kuhn sees the necessary conditions in the following factors: (1) the availability of conversion processes (2) the concern with engines and (3) the philosophy of nature. Needless to say, I fully agree with his third criterion, reject the second (dealt with in Chapter III), and I shall now briefly discuss the first. Additional conditions have been cited by me and dealt with above (like the double tradition in mechanics, the physiological background, and the fact that there is no direct connection between the theory of heat and any conservation ideas).

The conversion processes

Processes of conversion had been available for more than fifty years in 1850. Since the work of Sadi Carnot it was also clear that at least a great part of work could be turned into heat. The numerical measurements of Gay-Lussac, of Mayer on their basis, and of Joule, were published in the scientific journals. That these revelations were not heeded, stems from the fact that they did not embrace the problem in its whole generality. Before Helmholtz, none of the workers in the field who realized that there was a principle of conservation of energy, supported their argument by the power of rational proof. Most of them needed a certain readiness on the part of the reader to believe in this principle. There was certainly no equally lucid mathematical formulation before Helmholtz, and in the case of Colding and Mayer, there was a clear commitment to a 'conservation of force', whatever this force may have been. We shall soon see what was understood by the term in the conceptual framework of scientists like Faraday or Grove. On the other hand, a man like Joule, with distinct thought processes and performing exact measurements, never generalized his findings to include all phenomena known in physics, or dealt with the mathematically well-established mechanical principle of conservation of energy. In a paper by E. C. Watson[2] we find that in the writer's opinion the only place where Joule ever gave a 'General Exposition of the Principle of Conservation of Energy' was in his popular lecture at St Ann's Church Reading Room in Manchester on 28 April 1847. Its title is 'On Matter, Living Force and Heat'; there we find the following:

From these facts it is obvious that the force expended in setting a body in motion is carried by the body itself, and exists with it and in it, throughout the whole course of its motion. This force possessed by moving bodies is termed by mechanical philosophers *vis viva*, or *living force*. The term may be deemed by some inappropriate, inasmuch as there is no life, properly speaking, in question; but it is *useful* in order to distinguish the moving force from that which is stationary in its character, as the force of gravity. When, therefore, in the subsequent parts of this lecture I employ the term *living force*, you will understand that I simply mean the force of bodies in motion. The living force of bodies is regulated by their weight and by the velocity of their motion. We might reason, *a priori*, that such absolute destruction of living force cannot possibly take place, because it is manifestly absurd to suppose that the powers with which God has endowed matter can be destroyed any more than that they can be created by man's agency; but we are not left with this argument alone, decisive as it must be to every unprejudiced mind. The common experience of everyone teaches him that living force is not *destroyed* by the friction or collision of bodies. We have reason to believe that the manifestations of living force on our globe are, at the present time, as extensive as those which have existed at any time since its creation, or, at any rate, since the deluge—that the winds blow as strongly, and the torrents flow with equal impetuosity now, as at the remote period of 4000 or even 6000 years ago; and yet we are certain that, through that vast interval of time, the motions of the air and of the water have been incessantly obstructed and hindered by friction. We may conclude, then, with certainty, that these motions of air, and water, constituting living force, are not *annihilated* by friction. We lose sight of them, indeed, for a time; but we find them again reproduced. Were it not so, it is perfectly obvious that long ere this all nature would have come to a dead standstill. What, then, may we inquire, is the cause of this apparent anomaly? How comes it to pass that, though in almost all natural phenomena we witness the arrest of motion and the apparent destruction of living force, we find that no waste or loss of living force has actually occurred? Experiment has enabled us to answer these questions in a satisfactory manner; for it has shown that, wherever living force is apparently destroyed, an equivalent is produced which in process of time may be reconverted into living force. This equivalent is *heat*.

The general rule, then, is that wherever living force is *apparently* destroyed, whether by percussion, friction, or any similar means, an exact equivalent of heat is restored. The converse of this proposition is also true, namely, that heat cannot be lessened or absorbed without the production of living force, or its equivalent attraction through space. You see, therefore, that living force may be converted into heat, and that heat may be converted into living force, or its equivalent attraction through space. All three, therefore—namely, heat, living force, and attraction through space (to which I might also add *light*, were it consistent with the scope of the present lecture)—are mutually convertible into one another.

It is also interesting that Kelvin, in his well-known narrative on how he heard Joule lecturing, finds not only that Joule's results do not agree with those of Sadi Carnot (although the latter had been supposed by many to have been in possession of the principle of conservation of energy), but seemingly Joule did not even know about Carnot, for Kelvin had to inform him:

I can never forget the British Association at Oxford in the year 1847, when in one of the sections I heard a paper read by a very unassuming young man who betrayed no consciousness in his manner that he had a great idea to unfold. I was tremendously struck with the paper. I at first thought it could not be true because it was different from Carnot's theory, and immediately after the reading of the paper I had a few words of conversation with the author James Joule, which was the beginning of our forty years' acquaintance and friendship. On the evening of the same day that very valuable Institution of the British Association, its conversazione, gave us opportunity for a good hour's talk and discussion over all that either of us knew of thermodynamics. I gained ideas which had never entered my mind before, and I thought I too suggested something worthy of Joule's consideration when I told him of Carnot's theory.[3]

And again, in a letter to Bottomley:

I made Joule's acquaintance at the Oxford Meeting, and it quickly ripened into a life-long friendship. I heard his paper read at the sections, and felt strongly impelled to rise and say that it must be wrong, because the true mechanical value of heat given, suppose to warm water, must, for small differences of temperature, be proportional to the square of its quantity. I knew from Carnot's law that this must be true (and it is *true*); only now I call it 'motivity', in order not to clash with James Joule's 'Mechanical Value'. But as I listened on and on, I saw that (though Carnot had a vitally important truth not to be abandoned) Joule had certainly a great truth and a great discovery, and a most important measurement to bring forward.[4]

Let me repeat it once more, it is not the modern terminology which is missing, but the concept of conservation in all its generality in mathematical terms. Thus one could say, that those who did exact experimental work had no deep-seated conservation ideas or no mathematical approach, while those who were committed to a principle of conservation, believed in the conservation of some vague entity called 'force'. Not until Helmholtz showed it mathematically, by a kind of dimensional argument (it had to be commensurable with mv^2), did it become clear what was involved by talking about this force. Even then, it was sufficiently vague to elude the activity of Faraday's fertile mind and become the foundation of his 'field' concept.

I wish to deal with only two of the successful claimants to the title, not because they are more successful, or nearer the end, but because they are intrinsically more interesting from the philosophical point of view. However, before that, I hope to update the terminology to the 1840s.

More 'words'

We stopped in the sections on 'words' in Chapter II with the story of 'energy'. Since then, the term 'potential' has reappeared and also 'kinetic' has entered scientific language. *La Grande Encyclopédie* in 1765 says the following under the item 'potentiel':

Potentiel adj. (phys.) froid potentiel est un mot relatif par lequel on fait connaître qu'une certaine chose n'est pas actuellement froide au toucher, mais qu'elle l'est dans ses effets et ses opérations, lorsqu'on le prend intérieurement.

Fifty years later, having peeled off the last remnants of the Aristotelian ideas of motion, the useful concept of potentiality could again enter physical science by the front door. In 1828 G. Greene introduced the concept of 'potential function'. (The word 'potential' was also used somewhat earlier in physics, as in Euler's '*vis potentialis*' but in a less clearly defined way and it was never really accepted by the whole of the 'scientific community' of the time.) Greene's definition was:

Nearly all the attractive and repulsive forces in nature are such that if we consider any mathematical point p, the effect, in a given direction, of all the forces acting upon that point . . . will be expressed by the partial differential of a certain function of the coordinates which serves to define the point's position in space. The consideration of this function is of great importance in many inquiries . . . we shall often have occasion to speak of this function and will therefore for abridgement call it the *potential function* arising from the system S.[5]

In 1853 Rankine introduced the expression of 'potential energy' as against Thomson and Tait's 'actual energy' (a last reminder of the Aristotelian juxtaposition). Finally, from 1870 and later the concept of 'potential' is used without necessarily requiring a complement, and has a clearly defined physical meaning—as the work required to move a unit charge against a field of force.

The story of 'kinetic' is much shorter. The term 'kinetic energy' was introduced by Lord Kelvin in the 1870s. Here again one has to beware of translation. I previously gave an example of how Leibniz used the term 'energy', and another example from Leibniz appears below. In the Leibniz *Selections*, edited by Philip P. Wiener, there is the translation of a letter from Bayle, written in 1702, where we find:

It is by way of a new approach which has taught me that not only is the force conserved but also the very quantity of kinetic action, which is different from that of the motion.[6]

It is thus clear, that not only does the principle of conservation of '*vis viva*' originate with Leibniz, but so do the beginnings of its modern name. However, the original French of the letter uses the old, familiar concept, repeatedly discussed in the Cartesian–Leibnizian controversy: 'action motrice', and not 'action cinétique!'

Colding and Mayer

The two energy-conservationists I wish to mention briefly are Ludvig Colding and Julius Robert Mayer.

Colding was a Danish carpenter, who very early became acquainted with Ørsted, and on the latter's advice studied at the Copenhagen Arts Academy, became a journeyman and got acquainted with various machines while preparing drawings for the Royal Mint. Not much is known about him—he wrote few articles, his original 1843 paper, in which he announced the establishment of the conservation law no more generally than Joule, Mayer and many others, has not even fully been translated into English. My interest in him is because he acted as Faraday to Ørsted's Davy—with the considerable exception that Colding seems greatly to have been influenced by the philosophy of his teacher and master, and thus developed an almost mystical belief in the 'Naturkräfte'. Nevertheless, his approach was basically scientific, so much so, that Ørsted found his pupil's descriptions too 'exact' for his taste. But before quoting him in any detail, let me make clear how meagre my sources are—part of the original memoir in Danish,[7] the much later English article in the *Philosophical Magazine* (1871) and an article by Per F. Dahl.[8]

The 1871 article does not mean much, and was supplied by Professor Tait, in view of the intense controversy on priorities, mainly to oppose Tyndall's championing of Mayer. It is formulated in modern language (using the term 'energy') and applies the full mathematical apparatus first used for this purpose by Helmholtz twenty-five years before this communication. Thus, we cannot learn much here. Per F. Dahl says that Colding's first idea of conservation occurred to him in 1840, and that he was led to it by 'd'Alembert's principle about lost forces', and that the 'start for these speculations was just as much of a religious and metaphysical nature'. In his article 'On the History of the Principle of the Conservation of Energy'[9] Colding writes:

As the forces of nature are something spiritual and immaterial, entities whereof we are cognizant only by their mastery over nature, these entities must of course be very superior to everything material in the world; and as it is obvious that it is through them only that the wisdom we perceive and admire in nature expresses itself, these powers must evidently be in relationship to the spiritual, immaterial and intellectual power itself that guides nature in its progress; but if such is the case, it is consequently quite impossible to conceive of these forces as anything naturally mortal or perishable. Surely, therefore, the forces ought to be regarded as absolutely imperishable.

Surely, Ørsted too would have signed this manifesto. It was on Ørsted's advice that he delayed submitting his memoir until some experimental evidence was collected, from 1840 till 1843. The evidence was even then meagre, but at least Colding could quote several experiments by other people, performed for purposes different from his own, as evidence for his theory. The committee of the Royal Danish Academy gave the following opinion of the submitted thesis:

The main thought of this thesis by polytechnic candidate Colding on which the Academy has requested an opinion, is that the forces which are lost for mechanical action by friction, resistance, pressure etc. create internal actions in the bodies, for instance, heat, electricity, and the like, and that these are in proportion to the lost forces. To strengthen his claim he has carried out a series of experiments on the heat by friction.

We find both that his main thought fully deserves an experimental trial and that his experiments are as satisfactory as one might demand considering the available equipment. We suggest therefore to encourage him in the continuation of these experiments by granting him the required equipment which is estimated not to exceed the cost of 200 Rigsdaler.

The thesis itself was printed only in 1856. At that early date the formulation was even vaguer and certainly less exact. The main element of vagueness (if the translation can be trusted) is the usage of the term 'action' with all the associated overtones which, though they go back to Descartes and Newton, are much more connected with the theories of Leibniz.

When certain moving forces act on a material point, and these do not hold each other in equilibrium, then there appears a certain amount of motion corresponding to the resultant force. This amount of motion is transmitted to the surrounding material parts, and distributes itself further from these in the same manner indefinitely, such that the original transmitted amount of motion within a short time is distributed throughout such a large mass that all discernible trace of this action has disappeared.

But it does not appear to me that one may justifiably assume that an action can eventually disappear completely in the material without appearing in some manner in its original magnitude. It further seems to me inherent in the nature of things that the forces which sensibly disappear must again appear as active in some other way. This thought occurred to me a long time ago and I have not since been able to discard it. On the contrary I have become convinced to such a degree of certainty of the correctness of this statement that I will try to present it as a common law of nature: when a force sensibly disappears, it undergoes only a change, a form, and becomes thereupon active in other forms.[10]

From the purely scientific point of view there is not much more to the memoir as far as I can see. The question which remains to be answered for the case of Ørsted, Colding, Ohm, Seebeck, and many others is to what extent was their belief in a conservation law connected with their excellent experimental work in other fields, or was it a philosophical commitment which had not much to do with the rest? It seems to me that the answer will turn out to be that their philosophical commitment came first, and even became a motive for their research. In the case of Ørsted at least, this seems to be beyond doubt, but none of them were great enough to translate their beliefs into mathematico-physical terms directly, and draw all the implications, which could then be tested experimentally. Indeed the road was different; when Helmholtz came around to translate his commitment into those physico-mathematical terms, not only had he more mathematical knowledge, but also more experimental results were available to him to confirm him in the correctness of his conclusions.

Mayer is an altogether different case. He may be classed with those who, on the one hand, were committed philosophically to great conservation laws of nature, while on the other, inherited the old medico-physiological tradition which tended to accept these great laws in any case. His attempt to seem 'modern' as far as denunciation of the metaphysicians went is natural; this was very much the *bon ton* of his days. An important and rewarding study could be made on Mayer's work, but not within our framework: such a study ought to trace the development of his thought and to compare carefully and systematically all his published works, beginning with his earliest ideas on the subject (the letters written during his voyage to the Dutch colonies), and the later books and essays, including his very interesting early denunciation of the 'Wärmetod'.

It was not by accident that Julius Robert Mayer, the one who is mentioned most often together with Helmholtz as the discoverer of the conservation principle, arrived at his problem through physiology. True, most of this renown came to him through the campaign of priorities that many waged for his sake. But there is more to it.

J. R. Mayer was no mere 'hunch' philosopher. Though it is difficult to find the logical path from the change in the colour of venous blood in the tropics to a full annunciation of the principle of conservation of energy, it was nevertheless a full realization of the principle. The depth of his understanding is clearly illustrated by the fact, that in his 'Dynamik des Himmels', he was one of the first scientists to disclaim the validity of the 'Wärmetod'—which seemed in the first decades after Clausius's work on the entropy principle to be inviolable—and he based his opposition on the conservation of energy principle. There is thus no doubt as to the universality and depth of his genius. However, his work transmits the real flavour of that stage of development when the concepts were indeed in a state of flux, in mid-nineteenth-century biology. I will quote a passage from Mayer's first paper, in which he formulates for the first time his theory of the mechanical equivalent of heat. In all fairness to this much-abused genius, let me say in advance that though his first version was full of mistakes, these were corrected in the next, after enlisting the help of some of his physicist friends. The longer, corrected version appeared in 1845 and upon it his fame justly rests. However, for our purposes the first paper is much more useful as it shows in what muddled terms the principle of conservation emerged:

Der Zweck folgender Zeilen ist, die Beantwortung der Frage zu versuchen, was wir unter 'Kraeften' zu verstehen haben, und wie sich solche untereinander verhalten. Waehrend mit der Benennung Materie einem Objekte schon bestimmte Eigenschaften, als die der Schwere, der Raumerfuellung, zugeteilt werden, knuepft sich an die Bennennung Kraft vorzugsweise der Begriff des unbekannten, unerforschlichen, hypothetischen. Ein Versuch den Begriff von Kraft ebenso praezis als den von Materie aufzufassen, und damit nur Objekte wirklicher Forschung zu bezeichnen, duerfte mit den daraus fliessenden Konsequenzen, Freunden klarer Hypothesen freier Naturanschauung nicht unwillkommen sein. Kraette sind Ursachen, mithin findet auf dieselben volle Anwendung der Grundsatz: *causa aequat effectum*. Hat die Ursache c die Wirkung e so ist $c=e$; ist e wieder die Ursache einer andern Wirkung f, so ist $e=f$, usf. $c=e=f=\ldots=c$. In einer Kette von Ursachen and Wirkungen kann, wie aus der Natur einer Gleichung erhellt, nie ein Glied oder ein Theil eines Gliedes zu Null werden. Diese erste Eigenschaft aller Ursachen nennen wir ihre Unzerstörbarkeit.[11]

I shall use the English translation from a collection called *The Correlation and Conservation of Forces*—the Faraday-influenced title is evident:

The following pages are designed as an attempt to answer the questions, what are we to understand by 'Forces'? and how are different forces related to each other? Whereas the term *matter* implies the possession, by the object to which it is applied, of very definite properties, such as weight and extension; the term *force* conveys for the most part the idea of something unknown, unsearchable, and hypothetical. An attempt to render the notion of force equally exact with that of matter, and so to denote by it only objects of actual investigation, is one, which, with the consequences that flow from it, ought not to be unwelcome to those who desire that their views of nature may be clear and unencumbered by hypotheses . . . Forces are causes: accordingly, we may in relation to them make full application of the principle—*causa aequat effectum*. If the cause '*c*' has effect '*e*' then $c = e$; if, in its turn, '*e*' is the cause of a second effect '*f*', we have $e = f$ and so on: $c = e = f = \ldots = c$. In a chain of causes and effects, a term or part of a term, can never, as plainly appears from the nature of the equation, become equal to nothing. This first property of all causes we call their indestructibility.[12]

To underline his argument, Mayer sees the term 'force' as having been hopelessly muddled and unscientific. By announcing his conservation principle, by defining forces as causes, and by having recourse to an old scholastic argument—'causa aequat effectum' (though also used by Leibniz, but in much clearer terms)—Mayer hopes to have made the concept of 'force' well-defined and less 'hypothetical'! Nevertheless, out of this muddled terminology sprang a deep understanding of the principle in all its generality. What Mayer could not do was to give the principle its mathematical form, that is to define our concept of energy. At this stage an immediate qualification is necessary—the point I am trying to make is not that the conservation principle had to spring from a muddled terminology, but only that, as long as the concepts were in a state of flux, we cannot follow logically what happened between the stage when these vaguely formulated principles were conceived, and the point when the final, generalized form was perceived. While Mayer's conservation principle served him as a fruitful tool for further research, it was not yet in its final form—namely after the definition of the concept of energy—only achieved after the work of Helmholtz. As we shall see, most of the elements in Mayer's essay can also be found in Helmholtz's 'Über die Erhaltung der Kraft': namely, the Kantian matter-force dualism, a vague concept of what 'Kraft' is, and a deep conviction in its indestructibility.

Helmholtz: what he did

We have now arrived at Helmholtz's 'Erhaltung der Kraft'[13] itself. After a brief survey of what he actually did, most of my argument will consist of textual criticism, to show that the very choice of words on his part is such that the term 'Kraft' is ambiguous, and that this ambiguity is inherent in the problem. Let me repeat the argument from the introductory chapter, namely, that this is a case of 'concepts in flux' or of concept-creation— i.e. scientific activity at its highest level. Helmholtz started out with a vague, undefined 'Kraft', which he believed to be conserved in Nature, and to which all other physical forces were related—even the unknown 'vital force' which had been occupying him so much in the last two years —and this force had to be brought into mathematical relationship with the entity which rational mechanics has proved to be conserved. Moreover, that entity must somehow tie up with the Newtonian force concept, and there must be at least a clear relationship between the two. Needless to say, the principle of the impossibility of a *perpetuum mobile* must be a consequence of this conservation law, whether we deal with dead or live matter.

In the following pages I shall use the English translation by John Tyndall, and leave the discussion of this translation and of the problems of terminology for later. The essay is divided into an introduction and six sections. The introduction is mainly philosophical and we shall return to it in the appropriate chapter.

The first section covers the principle of the conservation of *vis viva*; presupposing the impossibility 'to produce force continually from nothing'. Then: 'Let us imagine a system of natural bodies occupying certain relative positions towards each other, operated upon by forces mutually exerted among themselves, and caused to move until another definite position is attained'. It is clear from the principle, that the 'quantity of work gained by this passage of the system from the first position to the second, and the quantity lost by the passage of the system from the second position back again to the first, are always equal'.

Otherwise a *perpetuum mobile* might be constructed. The mathematical expression of what has been attained is

$$\tfrac{1}{2}mv^2 = mgh.$$

This principle according to Helmholtz 'declares that when any number whatever of material points are set in motion, solely by such forces as they exert upon each other, or as are directed against fixed centres, the total sum of the *vires vivae* at all times when the points occupy the same relative position, is the same, whatever may have been their paths'. This principle applies only to central forces, as is well known. But Helmholtz now proves that 'in all actions of natural bodies upon each other, where the above principle is capable of general application, even to the ultimate particles of these bodies, such central forces must be regarded as the simplest fundamental ones'. Skipping the mathematical proof (it is very simple and straightforward) the conclusion is drawn: 'in systems to which the principle of the conservation of force can be applied in all its generality, the elementary forces of the material points must be central forces'.

In section 2 the Principle of Conservation of Force is dealt with. If φ is the intensity of the force which acts in direction r we take the three components of the force F as:

$$F_x = -\frac{x}{r}\varphi \qquad F_y = -\frac{y}{r}\varphi \qquad F_z = -\frac{z}{r}\varphi$$

and then from the second law of Newton and the conservation of *vis viva,* we obtain

$$\frac{1}{2}mQ^2 - \frac{1}{2}mq^2 \quad = -\int_r^R \varphi\,dr.^{14}$$

The important quantity is

$$\int_r^R \varphi\,dr$$

which will be called the sum of the tensions between the distances R and r, and the above law can be expressed as follows. 'The increase of *vis viva* of a material point during its motion under the influence of a central force is equal to the sum of the tensions which correspond to the alteration of its distance.' Now this is generalized to any number of points whatsoever. After this one can generally say that the 'sum of the existing tensions and *vires vivae* is always constant'. The relation of this law to the principle of virtual velocities is taken up next, and it is shown that this principle follows from that of the conservation of force. The laws of statics follow. The conclusions to this section are summed up under three headings:

(1) Whenever natural bodies act upon each other by attractive or repulsive forces, which are independent of time and velocity, the sum of their *vires vivae* and tensions must be constant; the maximum quantity of work which can be obtained is therefore a limited quantity.

(2) If, on the contrary, natural bodies are possessed of forces which depend upon time and velocity, or which act in other directions than the lines which unite each two separate material points, for example, rotatory forces, then combinations of such bodies would be possible in which force might be either lost or gained *ad infinitum.*

(3) In the case of the equilibrium of a system of bodies under the operation of central forces, the exterior and the interior forces must, each system for itself, be in equilibrium, if we suppose that the bodies of the system cannot be displaced, the whole system only being movable in regard to bodies which lie without it. A rigid system of such bodies can therefore never be set in motion by the action of its interior forces, but only by the operation of exterior forces. If, however, other than central forces had an existence, rigid combinations of natural bodies might be formed which could move of themselves without needing any relation whatever to other bodies.

Section 3 is the application of the principle in mechanical theorems. It is shown to apply to all motions which proceed under the influence of the general force of gravitation; to the transmission of motion through incompressible solid and fluid bodies, where neither friction nor concussion take place; to the motion of perfectly elastic, solid and fluid bodies. Among the elastic fluids one has to take into account 'waves on the surface of liquids, the motion of sound', and probably also of those of light and radiant heat. The only authority cited for this conclusion is Fresnel, who 'from the assumption that the motion of the limiting particles [in case of reflection or refraction between two media] in both media is the same, and from the conservation of *vis viva* deduced these laws of reflection and refraction.

Section 4 is on the force-equivalent of heat. Helmholtz lists the processes in which absolute loss of force had been assumed: these are the collision of inelastic bodies and friction. The explanation rests on the model of elastic solids where the *vis viva* which had previously been supposed lost absolutely, changes into tension, and all boils down to attraction and repulsion between elementary particles.

Joule's measurements are cited, but only the earlier ones, which had indeed been hardly reliable. Here, instead of accepting Joule's experimental work as proof, Helmholtz examines separately 'how far heat can correspond to an equivalent of force'. Helmholtz deals with the problem on both theories of heat, and his treatment of the caloric theory is well worth quoting:

Let us now turn to the further question, how far heat can correspond to an equivalent of force. The material theory of heat must necessarily assume the quantity of caloric to be constant; it can therefore develop mechanical forces only by its effort to expand itself. In this theory the force-equivalent of heat can only consist in the work produced by the heat in its passage from a warmer to a colder body; in this sense the problem has been treated by Carnot and Clapeyron, and all the consequences of the assumption, at least with gases and vapours, have been found corroborated.

To explain the heat developed by friction, the material theory must either assume that it is communicated by conduction as supposed by Henry, or that it is developed by the compression of the surfaces and of the particles rubbed away, as supposed by Berthollet. The first of these assumptions lacks all experimental proof; if it were true, then in the neighbourhood of the rubbed portions a cold proportionate to the intense heat often developed must be observed. The second assumption, without dwelling upon the altogether improbable magnitude of action, which according to it must be ascribed to the almost imperceptible compression of the hydrostatic balance, breaks down completely when it is applied to the friction of fluids, or to the experiments where wedges of iron have been rendered red-hot and soft by hammering and pieces of ice melted by friction; for here the softened iron and the water of the melted ice could not remain in a compressed condition.

Besides this, the development of heat by the motion of electricity proves that the quantity of heat can be actually increased. Passing by frictional and voltaic electricity—because it might here be suspected that, by some hidden relation of caloric to electricity, the former was transferred from the place where it was originated and deposited in the heated wire—two other ways of producing electric tensions by purely mechanic agencies in which heat does not at all appear, are still open to us, namely, by induction and by the motion of magnets. Suppose we possess a completely insulated body positively electric and which cannot part with its electricity; an insulated conductor brought near to it will show free $+E$, we can discharge this upon the interior coating of a battery and remove the conductor, which will then show $-E$; this latter can of course be discharged upon the exterior surface of the first or upon a second battery. By repeating this process, it is evident that we can charge a battery of any magnitude whatever as often as we please, and by means of its discharge can develop heat, which nowhere disappears. We shall, on the contrary, have consumed a certain amount of force, for at each removal of the negatively-charged conductor from the inducing body the attraction between both is to be overcome. This process is essentially carried out when the electrophorus is used to charge a Leyden jar. The same takes place in magneto-electric machines; as long as magnet and keeper are moved opposite to each other, electric currents are excited which develop heat in the connecting wire; and inasmuch as they constantly act in a sense contrary to the motion of the keeper, they destroy a certain amount of mechanical force. Here evidently heat ad infinitum may be developed by the bodies constituting the machine, while it nowhere disappears. That the magneto-electric current develops heat instead of cold, in the portion of the spiral directly under the influence of the magnet, Joule has endeavoured to prove experimentally. From these facts, it follows that the quantity of heat can be absolutely increased by mechanical forces, that therefore calorific phaenomena cannot be deduced from the hypothesis of a species of matter, the mere presence of which produces the phaenomena, but that they are to be referred to changes, to motions, either of a peculiar species of matter, or of the ponderable or imponderable bodies already known, for example of electricity of the luminiferous aether.

And the summing up of the problem is even clearer:

That which has been heretofore named the quantity of heat, would, according to this, be the expression, first, of the quantity of *vis viva* of the calorific motion, and secondly, of the quantity of those tensions between the atoms, which, by changing the arrangement of the latter, such a motion can develop. The first portion would correspond to that which has been heretofore called free heat, the second with that which has been named latent heat.[15]

Problems of heat radiation are then examined; following this the laws of Clapeyron and Holtzmann are dealt with. The conclusion is, that for gases the force-equivalent from Clapeyron's formula is equal to the result of Holtzmann, while 'its applicability to solid and liquid bodies remains as yet doubtful'.

Section 5 deals with the force equivalent of electric processes. Above we have seen how Helmholtz referred to the problems of heat in the electrical processes; here he distinguishes static electricity from galvanism. The conclusion is, that 'the increase of *vis viva* due to any motion whatever is equal to the excess of the potential at the end of the route over its value at the beginning'. It is explained how electrical tensions are converted to mechanical action and the principle is shown to apply to many special cases. The same is done for galvanism, where two kinds of conductors are distinguished—'(1) those which conduct in the manner of metals and follow the law of the tension series; (2) those which do not follow this law'. After careful explanations it is finally shown that the 'heat developed in any circuit whatever, where the conduction is effected through any number of branches, if the law of Lenz be true for fluid conductors, as found by Joule', will be given by

$$Q = I^2 Rt = nAIt$$

$t =$ time
$I =$ current intensity
$n =$ number of elements
$A =$ electromotive force
$R =$ resistance
$Q =$ total quantity of heat

(the Lenz law referred to here is the first part $Q = I^2 Rt$). The cells of Daniell and Grove are then examined, when the argument turns highly technical, and we shall not follow it here. Thermo-electric currents are also covered.

Section six concerns the force-equivalent of magnetism and electro-magnetism. The argument here begins with the statement, 'through the attractive and repulsive forces which a magnet exerts upon other magnets, or upon soft iron, it is capable of generating *vis viva*'. The assumption with which Helmholtz works is the two-fluid theory:

As the phenomena of magnetic attraction may be completely deduced from the assumption of two fluids which attract or repel in the inverse ratio of the square of the distance, it follows without going further than the deduction made at the commencement of this memoir (about central forces) that in the motions of magnetic bodies the conservation of force must take place.

These motions are now considered in detail. Electrodynamic phenomena are then dealt with on the basis of the Ampère model, which basically, again deals with attractive and repulsive forces acting on elements of current. These are developed in some technical detail, and the limitations of the theory at that stage stated are also set forth.

In the same section, but in a new part, Helmholtz arrives at the problem of organic forces. The conclusions are again worth quoting in full:

Of known natural processes those of organic existences are still to be considered. In plants the processes are chiefly chemical, and besides these a slight development of heat takes place, at least in some: but the principal fact is, that a vast quantity of chemical tensions is here stored up, the equivalent of which we again obtain as heat by the combustion of the plants. The only *vis viva* which we know to be absorbed in the accomplishment of this is that of the chemical solar rays; we are, however, totally at a loss for the means of comparing the force-equivalents which are thereby lost and gained. Animals present some points in this respect which we can lay hold of. These take in oxygen and the complicated oxidizable combinations which are generated by plants, and give back the same, for the most part burnt, as carbonic acid and water, but in part reduced to simpler combinations; hence they consume a certain quantity of chemical tensions, and generate in their place heat and mechanical force. As the latter compared with the quantity of heat represents but a small quantity of work, the question of the conservation of force is reduced to this, whether the combustion and metamorphosis of the substances which serve as nutriment generate a quantity of heat equal to that given out by animals. According to the experiments of Dulong and Despretz, this question can be approximately answered in the affirmative.

In conclusion I must refer to some remarks of Matteucci's which have been directed against the views advocated in this memoir, and which appear in the *Bib. Univ. de Genève*, No. 16, 1847, 15 May, p. 375. He proceeds from the proposition, that according to the above views a chemical process could not generate so much heat where it at the same time develops electricity, magnetism, or light, as when this is not the case. He takes pains to show by a series of measurements which he adduces, that zinc, during its solution in sulphuric acid, generates just as much heat where the solution is effected directly by chemical affinity as when it forms a circuit with platinum; and that an electric current produces just as much chemical and thermic action while it deflects a magnet as when no such deflection is produced. That Matteucci regards these facts as objections, is due to his total miscomprehension of the views which he undertakes to refute, which will be at once evident from a consideration of our statement of the subject. He then brings forward two calorimetric experiments on the heat which is developed by the combination of caustic baryta with concentrated or dilute sulphuric acid, and on that generated by the same electric current in a wire immersed in gases of different cooling capacities, whereby the above mass and the wire were something glowing and sometimes not. He finds the quantity of heat in the former cases not less than in the latter. When we, however, reflect upon the incompleteness of our calorimetric arrangements, it will not appear extraordinary that differences of cooling through radiation, which are due to the fact that this radiation, according to its luminous or non-luminous nature, passes with less or more difficulty through the surrounding diathermanous bodies, escape observation. In the first experiment of Matteucci the union of the baryta with sulphuric acid was effected in a non-diathermanous leaden vessel, where the luminous rays were completely prevented from escaping outwards. The imperfections of Matteucci's methods in carrying out these measurements need not be further dwelt upon.

By what has been laid down in the foregoing pages, I believe I have proved that the law in question does not contradict any known fact in natural science, but in a great number of cases is, on the contrary, corroborated in a striking manner. I have endeavoured to state in the most complete manner possible, the inferences which

flow from a combination of the law with other known laws of natural phaenomena, and which still await their experimental proof. The object of this investigation was to lay before physicists as fully as possible the theoretic and practical
importance of a law whose complete corroboration must be regarded as one of the
principal problems of the natural philosophy of the future.

Let me now look systematically at the assumptions underlying this
treatment: Helmholtz was very much committed—*a priori*—to two
fundamental beliefs: (a) that all phenomena in physics are reducible to
mechanical processes (no one who reads Helmholtz can doubt this), and
(b) that there must be some basic entity in Nature which is being conserved (this does not appear in so many words in Helmholtz's work and
will occupy us further on). His physiological studies made him a firm
believer in reductionism: i.e. all organic processes are reducible to
physics. If we collect these premises systematically the argument looks as
follows:

(i) Newtonian 'force' is a fundamental concept in mechanics.
(ii) Physics is reducible to mechanics.
(iii) The fundamental concept in physiology is 'force of life'; physiology is reducible to physics, i.e. to mechanics.
(iv) There is a basic entity in Nature which is being conserved.

Conclusion: The basic entity which is being conserved must be 'Kraft'.

But for the final formulation we have to add another premise:

(v) The Lagrangian formulation of mechanics is equivalent to the
Newtonian formulation mathematically and conceptually: The
Lagrangian formulation has as its fundamental entity 'kinetic
energy plus potential energy'; this fundamental sum is being conserved.

Conclusion: The basic entity 'Kraft' which is being conserved in Nature
must be equivalent in dimension and form to mechanical energy. This is
the generalized conservation of energy principle.

It is exactly this programme which was executed by Helmholtz in his
paper 'Über die Erhaltung der Kraft'. After that, the concept of energy
underwent the fixing stage; the German 'Kraft' came to mean simply
'energy' (in the conservation context) and later gave place slowly to the
expression 'Energie'. The Newtonian 'Kraft' with its dimensions of mass
times acceleration became simply our 'force'.

This systematization requires a few qualifying statements.

(a) Such an analysis can naturally be done only *ex post facto*. The
separation of the two conclusions presupposes an awareness of the

difference between the conserved 'Kraft' and the later 'Energie' which again presupposes that the concept has passed the state when it was 'in flux', i.e. it presupposes the final conclusion.

(b) I did not refer to the impossibility of perpetual motion as one of Helmholtz's premises, since I do not think that it occupied an important place among his presuppositions, in spite of what he says; but this has already been dealt with in the chapter on mechanics. Naturally, Helmholtz knew from early on, that a *perpetuum mobile* was not possible. This had been clear for almost a hundred years, without having resulted in an immediate, precise formulation of the principle of conservation of energy in all its generality.[16] For Helmholtz the true meaning of this impossibility was that any conservation law must imply the impossibility of perpetual motion.

(c) If one still feels inclined to ask why Helmholtz did not plainly state after 1847, that from then on, he and everyone else, had to distinguish clearly between the concepts of force and energy, the answer simply is, that he was not aware that there had ever been any confusion. In creating a new discovery or a new concept one has to think in, or work with, a set of other concepts: the logical connection between these two sets cannot be formulated in the same conceptual framework, but outside or beyond it. This is one sense in which one can understand the thesis, that a 'logic of discovery' is not possible.[17] The set of concepts before the discovery included '*vis viva*', 'Spannkraft', 'living forces', (Kraft), etc., not as instances of the one entity which is being conserved.

(d) In chapter II, on mechanical traditions, we dealt with Poincaré and his *Science and Hypothesis*. Though in a different context, he shows what happens if this historical point is forgotten. He considers the principle as an experimental law, and formulates it as follows: 'The sum of the kinetic energy, the potential energy and the energy of state is conserved.' If after that he wants to know what is the energy of a system where the contributions of forces dependent on position, velocity or state are not separable into different terms, he has no other answer but: '"There is something which remains constant" and this will be defined as energy.'[18]

Helmholtz v. Faraday

For contrast, let us now examine the case where a great scientist is committed to a conservation principle in Nature, without the guidelines of Lagrangian mechanics. Naturally, I mean Michael Faraday. He was brought up, or rather re-educated himself, on Newtonian mechanics, and for lack of mathematical background among other reasons he was not acquainted with the Euler–Lagrange formulation of analytical mechanics.

Consequently, his quest for a great unifying principle of Nature must have centred around the concept of force. After the investigations of L. Pearce Williams on Faraday,[19] I do not have to convince anyone as to how deeply Faraday was committed to a metaphysical belief in the 'Conservation of Force'. And in this case at least, there is no question whatsoever whether Faraday used 'force' but meant 'energy', or whether he really meant 'force'. Pearce Williams writes:

It must be first stated strongly that in this essay Faraday was definitely not dealing with the conservation of energy or the conservation of momentum, but with the conservation of force as he understood it.[20]

The essay mentioned is Faraday's famous 'On the Conservation of Force', published in 1857, ten years after Helmholtz's paper and several years after the latter became available in English, in John Tyndall's translation. There we find:

The case of a force simply removed or suspended, without a transferred exertion in some other direction, appears to me absolutely impossible. . . . The principle of conservation of force would lead us to assume that when A and B attract each other less, because of increasing distance, then some other exertion of power, either with or without them is proportionately growing. . . . When the equivalents of various forms of forces, as far as they are known, are considered, their differences appear very great; thus a grain of water is known to have electric relations equivalent to a very powerful flash of lightning. It may therefore be supposed that a very large apparent amount of force causing the phenomena of gravitation may be the equivalent of a very small change in some unknown condition of the bodies whose attraction is varying by change of distance.[21]

These passages illustrate beyond doubt that the entity dealt with is not energy; Faraday's peculiar 'force' is directional, and is equivalent to some 'extension of power'. To understand what he is driving at, one must keep in mind his definition of force (about which more later): 'a source of all possible actions of the particles, or materials of the universe'. Faraday knew of Helmholtz's paper, and he must have been impressed much more by the written text than by the mathematical treatment. If one compares his use of the word 'force' with Helmholtz's use of 'Kraft', the similarity is obvious. There is a letter by Maxwell to Faraday which Pearce Williams published in his Faraday book for the first time, and which sheds light on the whole problem. Faraday had asked Maxwell what he thought of his essay 'On the Conservation of Force'. To this Maxwell answered in a letter, where he says:

Now first I am sorry that we do not keep our words for distinct things more distinct and speak of the 'conservation of Energy' as applied to relations between the amounts of '*vis viva*' and of 'tensions' in the world.[22]

Let me interject here, that if anyone has taken Maxwell's version of Helmholtz's use of the words 'Kraft' very seriously (see above) here we have another proof. Maxwell's tendency to attribute to great scientists only 'correct' opinions is so strong than even he assumed that Faraday has distinct concepts and he is just not careful enough with his choice of words; in the same letter he defines force and energy very clearly:

Energy is the power a thing has of doing work arising either from its own motion or from the tension subsisting between it and other things. Force is the tendency of a body to pass from one place to another and depends upon the amount of change of 'tension' which that passage would produce.[23]

But with this definition Faraday disagreed basically; this was not his conception of force:

I perceive that I do not use the word 'force' as you define it 'the tendency of a body to pass from one place to another'. What I mean by the word *is the source or sources of all possible actions of the particles or materials of the universe.*[24]

The italics are mine. Needless to say Faraday never proved, or even tried to prove, his principle. He accepted his principle in an *a priori* fashion; had he never felt the inclination to translate every physical law into mathematical language—as was the case with Helmholtz—and had he possessed the required mathematical training, he would either have realized that 'force' is not being conserved, or would have discovered the principle of the conservation of energy. But this is idle speculation. As it was, his ideas were in a state of flux, his conservation principle was very vague and metaphysical, and out of this conceptual pool, grew his brilliant and physically fruitful idea of tubes of force, which finally led to one of the most fundamental concepts of modern physics: a well-defined, mathematically expressible 'field'.

'Kraft': textual criticism

Now to the textual criticism of 'Erhaltung der Kraft'. The concepts used in this work are: 'Arbeitskraft', 'Bewegungskraft', 'bewegende Kraft', 'lebendige Kraft', and 'wirkende Kraft'. But before testing whether one can or cannot simply replace these concepts by 'force' in some places and by 'energy' elsewhere—a procedure sanctified by the authority of Maxwell—there is a question which we have to ask. Why should Helmholtz, one of the most didactic writers, hesitate to create a new word for a 'clearly defined' concept if he was aware of any ambiguity? The usages of his time certainly did not prevent his coining the word 'Spannkraft' for the clearly defined mechanical entity which we call potential energy; he also created the concept of 'circulation' in his hydrodynamical vortex theory, and the expression 'cyclic variables' originated with him. Moreover, the word 'Energie' as we have seen above, was known to him and used by him; besides, as seen from various of his works, Helmholtz knew Young's work and had read the *Lectures on Natural Philosophy*, where the word had been introduced for our kinetic energy.

Furthermore, from the late 1870s, Helmholtz begins to use 'Konstanz der Energie', instead of 'Erhaltung der Kraft'. It is especially noteworthy, that in his 1881 notes to the original 1847 paper 'On the Conservation ot Force' (prepared for the edition of his collected scientific papers)[25] he again uses the new expression. In his 1887 article 'On the Principle of least Action', where a short review of the energy principle is also given, the old expression never occurs. Helmholtz was famous for his scrupulous honesty and for acknowledging errors and priorities. Even in the case of J. R. Mayer, where some authors tried to show that Helmholtz had ignored his contributions, Helmholtz did his utmost in later publications to repair the damage he had done, and to show Mayer's real worth. Had he thought in 1881 that his 1847 paper had been erroneous in any respect, or even merely confusing, he would certainly have mentioned the fact in these notes. The only explanation I can see for Helmholtz's not having done so, is that he himself saw the change in terminology, from 'Erhaltung der Kraft' to 'Constanz der Energie', as merely verbal. He was simply not aware of the conceptual changes and the fixing of the concept of energy. Helmholtz certainly did not think that anything had been wrong with his concept of 'Kraft' prior to his own proof of the conservation of it; in other words he was not aware that his concepts had been in a state of flux, and that the fixing of the concept of energy in his own mind and in the minds of his contemporaries, was due to his own proof of the conservation of energy in 1847. Psychologically this is clear: he could not have been aware of that.

Several passages are quoted below, each in the original and in translation, each passage having been numbered for easy identification when referred to in the comments.

(1) Alle Wirkungen in der Natur zurueckzufuehren seien auf anziehende und abstossende Kraefte, deren Intensitaet nur von der Entfernung der aufeinander wirkenden Punkte abhaengt.

... that all actions in nature can be ultimately referred to attractive or repulsive 'Kraefte', the intensity of which depends solely upon the distances between the points by which the 'Kraefte' are exerted.

(2) ... wenn wir von verschiedenartigen Materien sprechen, so setzen wir ihre Verschiedenheit immer nur in die Verschiedenheit ihrer Wirkungen, d.h. in ihre Kraefte.

... for when we speak of different kinds of matter we refer to difference of action, that is to differences in the forces of matter.

(3) Es ist eindeutend dass die Begriffe von Materie und Kraft in der Anwendung auf die Natur nie getrennt werden duerfen.

It is evident that in the application of the ideas of matter and 'Kraft' to nature, the two should never be separated.

(4) Eine reine Kraft waere etwas das dasein sollte und doch wieder nicht dasein, weil wir das Daseiende Materie nennen. Ebenso fehlerhaft ist es die Materie fuer etwas wirkliches, die Kraft fuer einen blosen Begriff erklaeren zu wollen, dem nichts wirkliches entspraeche. . . . Wir koennen ja die Materie eben nur durch ihre Kraefte, nie an sich selbst wahrnehmen.

A pure 'Kraft' would be something which must have a basis, and yet which has no basis, for the basis we name matter. It would be just as erroneous to define matter as something which has an actual existence, while force as a mere idea which has no corresponding reality. . . . Matter is only discernible by its 'kraefte', and not by itself.

(5) . . . die aeusseren Verhaeltnisse, durch welche die Wirkung der Kraeft modificirt wird, koennen nur noch raeumliche sein, also die Kraefte nur Bewegungskraefte . . .

. . . the only alteration possible to such a system is an alteration of position, that is of motion; the 'Kraefte' can only be 'Bewegungskraefte' . . .

So far everything seems extremely simple. In the first five passages we could probably have substituted 'forces' for 'Kraefte'. Yet, even here, the resulting expressions like 'forces of matter' and 'forces of motion' would be deceptive, because although they sound quite impeccable in English, they have no clearly defined meaning. It is interesting that the influence of the nineteenth century is still strong enough in our language for these expressions to sound good. If we examine their meaning carefully, we soon come to realize that they convey meaning only if force means 'a source of all possible actions of the particles or materials of the universe', i.e. Faraday's definition. But let us continue:

(6) Bewegungskraft . . . ist also zu definieren als das Bestreben zweier Massen ihre gegenseitige Lage zu wechseln. Die Kraft aber, welche zwei ganze Massen gegen einander ausueben, muss aufgeloest werden in die Kraefte aller ihrer Theile gegeneinander.

'Bewegungskraft' . . . is therefore to be defined as the endeavour of two masses to alter their relative position. But the 'Kraft' which two masses exert upon each other must be resolved into those exerted by all their particles on each other.

(7) Eine Bewegungskraft welche sie gegen einander ausueben . . .

A 'Bewegungskraft' therefore, exerted by each upon the other.

(8) Die Kraefte welche zwei Massen aufeinander ausueben, muessen nothwendig ihrer Grösse und ihrer Richtung nach bestimmt sein . . .

The 'Kraefte' which two masses exert upon each other must necessarily be determined according to their intensity and their direction . . .[26]

That the translation was done by Tyndall is noteworthy, because Tyndall was not only a well-known physicist, but held very similar ideas to Faraday on the nature of 'Forces'. It cannot be an accident that he felt the strong intellectual stimulus in Helmholtz's work, and he mentions him often. Again, it was Helmholtz who wrote the introduction to the German edition to Tyndall's *Faraday as a Discoverer*. Concerning the translation some comment is called for. Passage (8) is not fully translated; the missing sentences make it very clear in the German original, that these 'Kraefte' have both intensity and direction; the above passage (8) appears in my translation. Besides, the concept 'Bewegungskrafte' is once translated as moving force, whatever this may mean, and at other places as 'force, which originates motion' which sounds more like an explanation than a translation. Could it be that in 1854 when the usage of the word 'energy' in mechanics was gaining ground through the publication of Thomson and Tait's book, the translator felt somewhat uneasy? 'Bewegungskraft' for Helmholtz is a 'tendency', or as Tyndall puts it an 'endeavour';[27] a 'tendency' exerted by two masses having both intensity and direction. But this is not all:

(9) Es bestimmt sich also endlich die Aufgabe der Physikalischen Naturwissenschaften dahin, die Naturerscheinungen zurueckzufuehren an unveränderliche, anziehende und abstossende Kraefte. . . . Die Loesbarkeit dieser Aufgabe ist zugleich die Bedingung der vollstaendigen Begreiflichkeit der Natur. Die rechnende Mechanik hat bis jetzt die Beschraenkung fuer einmal weil sie ueber den Ursprung ihrer Grundsaetze nicht klar war . . .

Finally, therefore, we discover the problem of physical natural science to be, to refer natural phenomena back to unchangeable attractive and repulsive forces. . . . The solvability of this problem is the condition of the complete comprehensibility of Nature. In mechanical calculations this limitation of the idea of 'Bewegungskraft' has not yet been assumed . . .

The last sentence of the quoted German passage is missing in the translation. It should read: 'because it has not yet been clear about its own theorems'—but this is the crucial point here; what were those fundamental theorems or principles which had been not clearly established in mechanics at this stage? To this Helmholtz gives the answer at the beginning of the next section of his essay:

(10) Wir gehen aus von der Annahme, dass es unmoeglich sei, durch irgend eine Combination von Naturkoerpern bewegende Kraft aus nichts zu erschaffen.

We will set out with the assumption that it is impossible by any combination whatever of natural bodies, to produce 'bewegende Kraft' [here Tyndall uses 'force', Y.E.] continually from nothing.

The way I understand Helmholtz's approach to the principle of the impossibility of a perpetual machine—i.e. as a necessary implication of the basic conservation law—once it is proven, is that nothing is more natural than to treat this principle as one of 'those fundamental theorems' the exact status of which has not yet clearly been established in mechanics. On the other hand, this could sound offensive to the ears of English inductivists in 1854, and could thus have been left out on purpose! In the last passage quoted, the translator has given up the battle with the 'Bewegungskraft' too; he simply translates it now as 'force'. In passage (9) we have seen that 'Bewegungskraft' must be nearer to our 'force' than to our 'energy', having both intensity and direction. Could we reasonably say that the same word in passage (10) should be translated as 'energy' to sound more correct in modern times? The very formulation of the principle of the impossibility of a *perpetuum mobile* uses the vague term 'bewegende Kraft' and is thus difficult to translate exactly; but in the German original it all sounds very smooth. Even if we did translate 'Kraft' as 'Energie', how could we possibly talk about 'bewegende Energie'? This, even more than 'Bewegungskraft' sounds much more like an active entity: 'bewegende' should mean something active and having direction, and it is not exactly the same as 'Kraft der Bewegung'. (This could possibly be translated as 'energy of motion'). But the greatest difficulty with this passage occurs in the sentence immediately after that last quoted:

(11) Aus diesem Satze haben schon Carnot und Clapeyron eine Reihe theils bekannter, theils noch nicht nachgewiesener Gesetze . . . theoretisch hergeleitet.

By this proposition Carnot and Clapeyron have deduced theoretically a series of laws, part of which are proved by experiment, and part not yet submitted to this test. . . .

But what Carnot was talking about were '*force vive*' and '*force vive virtuelle*' both clearly defined terms with applicability limited to conservative systems; Clapeyron and Carnot both used the expression 'puissance motrice'. Whatever the difficulties in the case of Clapeyron or even more of Carnot (whether it is quantity of heat or entropy which they make to be conserved) may be (see above), they certainly did not have the 'force' *v.* 'energy' difficulty, and in this Helmholtz should have read them as if they were writing about energy—had this concept already been fixed in its meaning for him. Two more examples before quitting this pedestrian textual criticism:

(12) Denken wir uns ein System von Naturkoerpern, welche in gewissen raeumlichen Verhaeltnissen zueinander stehen und unter dem Einfluss ihrer gegenseitigen Kraeft in Bewegung gerathen, bis sie in bestimmte andere Lagen gekommen sind: so koennen wir ihre gewonnenen Geschwindigkeiten als eine gewisse mechanische Arbeit betrachten, und in solche verwandeln. Wollen wir nun dieselben Kraefte zum zweiten Male wirksam werden lassen, um dieselbe Arbeit noch einmal zu gewinnen, so muessen wir die Koerper...

Let us imagine a system of natural bodies occupying certain relative positions towards each other, operated upon by 'Kraefte' mutually exerted among themselves, and caused to move until another position is attained: we can regard the velocities thus acquired as a certain mechanical work and translate them into such. If we now wish the same 'Kraefte' to act a second time, so as to produce the same quantity of work, we must in some way, by means of other 'Kraefte' placed at our disposal, bring the bodies . . .

Should we now say that velocities are mechanical work? And finally:

(13) Nennen wir nun die Kraefte, welche den Punkt zu bewegen streben so lange sie eben noch nicht Bewegung erreicht haben? Spannkraefte, im Gegensatz zu dem, wasman in Mechanik lebendige Kraeft nennt, so wuerden wir die Groesse $\int_r R \gamma dr$ als die *Summe der Spannkraefte* zwischen die Entfernungen R und r bezeichnen koennen.

Calling the 'Kraefte' which tend to move the point m before the motion has actually taken place, 'Spannkraefte', in opposition to that what is meant in mechanics *'vis viva'*, then the quantity $\int_r R \gamma dr$ would be the *sum of the 'Spannkraefte'* between the distances R and r.

All this sounds very clear in German. The English translation with 'forces' for 'Kraefte' and 'tensions' or 'potential energy' for 'Spannkraefte' is also without difficulty at first glance. But let us remember that in passages (6) and (8) we found that 'Kraft' in this context, where it constitutes a 'tendency', should be rendered as 'force' (in our sense) and here these are 'Spannkraefte', which, as is clearly seen from the mathematical expression, must be understood as our energy, or rather our potential energy.

An unending succession of such passages could have been reproduced here. To sum up the argument—these passages, taken all together, are intelligible only if we leave the 'Kraft' unchanged: the Helmholtzian 'Kraft' is a vague entity, just that something which must be conserved in Nature: 'Kraft' is a concept in flux, a concept that will become fixed as our 'energy' only after Helmholtz has completed his mathematical reduction of all the 'Kraefte' to the form of mechanical energy.

It is interesting to note that Planck,[28] who, by the way, also reads Helmholtz 'correctly', i.e. attributes the use of the word 'Kraft' only to the different terminology of an earlier period, has the following to say:

So lange man also mit dem Worte Kraft keine klare Vorstellung verband, war ein Streit ueber das Mass der Kraft vollstaendig gegenstandslos. Indessen ist nicht zu verkennen dass dem besprochenen Streite dennoch ein wesentlich tiefer Inhalt zugrunde lag: denn die Parteien waren sich, wenn dies auch nur gelegentlich ausgesprochen wurde, in der Tat bis zu einem gewissen Grade ueber das einig, was sie unter 'Kraft' verstehen wollten. Descartes, sowohl als Leibniz hatten sicherlich eine, wenn auch nicht ganz klare praezise Vorstellung von der Existenz eines Prinzips, welches die Unverenderlichkeit und Unzerstoerbarkeit desjenigen ausspricht, aus dem alle Bewegung und Wirkung in der Welt hervorgeht.

As long as there was no clear notion connected with the word 'Kraft' any dispute on the quantity of this 'Kraft' was without a proper theme. Yet it must be admitted that this dispute had a much deeper content at its foundation; for, the parties to the dispute were to some extent united, even if they did not express this very clearly and often, as to what they wanted to understand under the word 'Kraft'. Descartes as well as Leibniz, had certainly some, even if not very precise, notion about a principle, which expresses the unchangeability and indestructibility of that from which all motion and action in the world emanates.

Could this statement merely be a different formulation of my claim, that new ideas and discoveries can and do often grow out of a pool of vague concepts? And if so, why does Planck apply it to Descartes and Leibniz, and at the same time ascribe to Helmholtz the concept of energy before the proof of its conservation? For Planck himself wrote that the concept of energy has its meaning only through the principle of its conservation. The reason, I believe, is that such an admission is much easier with respect to scientists of the past, whose works have either long been forgotten or whose concepts have long been fixed. Besides, it is easier to attribute vague scientific concepts to great philosophers who, however, were not equally successful as scientists. Finally, Planck wrote this in 1887, when Helmholtz was alive and active, the last breath of Naturphilosophie was still fluttering in the air, and its principles were still being fought. But, in the case of Descartes and Leibniz, what Planck does here, is not much different from a theory of concepts in flux, and constitutes a clear statement, that it is only after the solution of the physical problems in mathematical terms that the concepts became really fixed.

The reaction of the contemporaries

How did the contemporaries of Helmholtz and Faraday react to a 'principle of the conservation of force'? No, it is not fair to put Faraday and Helmholtz in a parallel situation in this context! Helmholtz's essay was technical, meant for and read before physicists. These physicists in some cases detected immediately the *a priori* flavour of the work, and in view of their strong inductivist philosophy, discarded it completely. Alternatively, they understood it immediately in all its generality, due to the mathematical treatment, and for them the concept of energy became immediately fixed; with no more problem about what Helmholtz actually had in mind. To this second group belonged two brilliant mathematicians: Jacobi and Du Bois-Reymond. Besides being a mathematician, the second was also a great physiologist; he had worked with Helmholtz in Johannes Müller's laboratory on the problem of physiological heat and understood immediately what Helmholtz's work meant for that problem. It is clear from Du Bois-Reymond's scattered remarks that he too realized the non-empirical basis of Helmholtz's work, but it did not deter his seeing its implications. Was this because he was a biologist? As to Jacobi, whether he saw it or not, I do not know. However, is it a mere accident, that he was the man who contributed to the final formulation of mechanics with special attention to the Principle of Least Action? In the meantime, the majority of physicists ignored Helmholtz's work (though not for long). Poggendorff, whose *Annalen* had rejected Mayer's work a few years previously, had also rejected the 'Über die Erhaltung der Kraft' and Helmholtz had to publish it privately. One must admit, that if one considered as the major task of science to eliminate all non-empirical elements from physics, or rather from the whole of natural philosophy, then it is small wonder that Mayer's and Helmholtz's work seemed too 'speculative'. The same happened in England with Faraday's work; we have already seen that Maxwell, who had great regard for Faraday, simply hoped to correct Faraday's concept of 'force', as if it had been a minor confusion. But Faraday's essay was not technical, and was read by many laymen and amateur scientists. Especially revealing is a debate between John Moore and the well-known physicist Charles Brooke, which was conducted in the pages of *Nature*. John Moore had written an article, 'The Heresies of Science', to which Brooke wrote a long reply, 'Conservation of Energy, a Fact—not a Heresy of Science'. Here Brooke says:

There is probably no term employed in physics that has been more misapplied, and in its misuse has led to greater confusion of ideas than 'force'.[29]

He then proceeds to define the concepts of 'force', 'energy', 'potential', 'sound', 'light' and 'heat'. Stating that he accepts Faraday's definition of force, which is (as already quoted twice) 'source or sources of all possible actions on the particles or materials of the universe', he considers his own definition just as an amplification, or rather an explanation of Faraday's:

Force is a mutual action between the atoms or molecules of matter, by which they are either attracted towards or repelled from, each other; and by this action energy is imparted to the matter put in motion. It may be further remarked that force is essentially either an attraction or a repulsion.

That for Brooke there were no vague concepts involved, and that he knew exactly what he was talking about (as did all those physicists, whose ideas were rooted in mechanics with its clear definition of kinetic and potential energies and the proof of conservation of the mechanical energy, and who did not seek to 'speculate' about the great unifying 'force of Nature') may be seen from the following passage:

If the above definition of force and energy be accepted, it is obvious that the term 'force', as used by Grove, Tyndall and many others, means sometimes force and sometimes energy.

Brooke is another scientist to whom the ideas are so clear, that he could not possibly imagine that someone of Faraday's stature would think in terms of a vague concept. He reads him, exactly as we saw that Maxwell did, 'correctly', i.e. 'knowing' what, in his opinion, Faraday should have meant. But John Moore had also read Faraday, and in his first paper we find:

A given motion viewed as a cause is force, while the very same motion thought of as an effect is energy.[30]

And then, in his reply to Brooke's sharp criticism of his first paper, Moore writes:

Mr Brooke says that 'energy' is the power of doing work. He does not tell us what he means by work. If he means motion in any of its modes, then he confounds what he holds to be distinct realities viz. force and energy. The 'theory of conservation of energy' as now maintained by physicists is opposed in several respects to the doctrine of the 'conservation of force' as held by Faraday—Stewart Brooke and others teach most explicitly that energy is not only constantly changing its form, but always shifting about from one position of matter to another. If I mistake not, Faraday asserts the very opposite respecting force. He seems to teach that each material particle, into whatever combinations it may enter, retains all its original forces.

'Stewart Brooke', well-known author of a physics textbook, and 'the others', are all those clear-thinking physicists who in the Thomson–Tait tradition, had very clear ideas about force and energy in mechanics, and no particular concern with the great unifying principles of Nature in philosophical terms. Moore quotes again Faraday's famous 'Conservation and Correlation of Forces' (we indeed wonder who had read Faraday in the spirit he meant to be read); he says:

Holding as I do, that forces are both conserved and correlated, I feel no difficulty whatever in accepting the facts established by Dr Joule. . . . But the molecular actions or motions are the effects of force but not the force itself. In no instant whatever can force be resolved into molecular action.

The last sentence is again a reflection of Faraday's definition of force. In short, those physicists who were dealing with specific problems, who were in these times mostly working in 'energetics', and who had very clear ideas about energy and force, either ignored Faraday's work as 'speculative' or simply read him 'correctly'. Some others whose questions were of the same kind and breadth as Faraday's understood him on a vague metaphysical level, and utter confusion was the result.

Holding with Meyerson[31] that the depth of understanding new ideas is best judged by seeing what the non-professionals have to say about it, let us look at Mr Nicolson's paper on the same topic as it appeared in *Nature* in 1871. He claims to have spent much thought on Faraday's 'On the Conservation of Force' and came to some conclusions about it:

What then does the 'conservation of forces' doctrine amount to in plain English? It amounts to the simple admission that the tendency to move is a property of matter inseparable from it and coexistent with it, and it is this tendency to move which is the cause of all the changes which we observe around us. There is however nothing new under the sun, for the old doctrine of Argan in *Le Malade Imaginaire* is revived again; when Argan answers his examiner for a license to practise in medicine, he says:

> Mihi a docto Doctore
> Domandatur causum et rationen quare
> Opium facit dormire
> Quia est in eo
> Virtus dormitiva
> Cujus est natura
> Sensus assoupire.

Many a clever student has laughed at this answer who little thought that research and experience would confirm it so strongly as they do now.[32]

I think that this passage speaks for itself. We have come to such a state where Faraday is thought to have revived the old scholastic occult qualities which had been so much attacked already in Newton's days! I mentioned above the different nature and impact of Helmholtz's work from those of Faraday's, but I chose to treat the reaction to Faraday's paper rather than the debate around Helmholtz's for two additional reasons. The first is because Helmholtz was very reluctant to participate in any scientific debate if he thought it fruitless. In his preface to the German edition of Thomson and Tait's *Natural Philosophy*, Helmholtz wrote:

I have as a rule considered it necessary to reply to criticisms of scientific propositions and principles only when new facts were to be brought forward or misunderstandings to be cleared up, in the expectation that, when all data have been given, those familiar with the science will ultimately see how to form judgement even without the discursive pleadings and sophistical arts of the contrary parts.[33]

Obviously, after having formulated his principle in all its generality and his own concepts having become fixed he did not think that there was anything to be cleared up. Whenever somebody did show confusion, Helmholtz classified him as one not sufficiently familiar with the science. He was rather astonished that when he published his 1847 paper (considering it only a few formulations and generalizations of well-known facts), that many physicists classified him as a 'speculative philosopher' and tried to prove him wrong. But he did not enter into discussion with them, but relied on the good judgement of his friends, Dubois-Reymond and the famous mathematician Jacobi, who supported his cause.

My other reason for dealing with the debate around Faraday was that Helmholtz himself participated in it, and was preoccupied with Faraday's achievements. When Tyndall wrote his *Faraday as a Discoverer* it was immediately translated into German and, as mentioned above, Helmholtz wrote the preface to it. I will quote from the English translation of this preface as it appeared in *Nature*. Helmholtz is talking mainly about Faraday's lines of force:

It is moreover by no means for the philosopher only that such an insight possesses interest. His interest certainly is the most immediate, for it had hardly been the lot of any single man to make a series of discoveries so great and so pregnant with the weightiest of consequences as those of Faraday. Most of them burst upon the world as surprises, the products apparently of an inconceivable instinct; and Faraday himself, even subsequently was hardly able to describe in clear terms the intellectual combinations which led to them. These discoveries, moreover, were all of a kin calculated to influence in the profoundest manner our notions of the nature of force. In the presence of Faraday's electromagnetic and diamagnetic discoveries more particularly, it was impossible for the old notions of force acting at a distance to maintain themselves without submitting to essential expansions and alterations.[34]

This is a very revealing passage indeed. Like Planck, when writing about Descartes and Leibniz (see above), when he was not trying to convince himself and others about the anti-hypothetical, purely empirical, character of good science. Helmholtz here acknowledges fully the vague, unformulated ways in which Faraday's mind worked. It could be that he thought this to have been a peculiarity of Faraday, but he certainly admits how very much those hypothetical, unproved ideas did result in a change of our conception of force. Did he imagine that Faraday had a clearly defined idea of force, a new one, and went out to make his discoveries with it? Faraday's own definition of force would disprove this. In my eyes this is an indirect admission by Helmholtz himself, that indeed while Faraday's discoveries were being made his concepts were vague. But a few passages later Helmholtz, probably not having realized the implications of his praise, goes on to reprimand Faraday for his 'misapprehension' of the law of conservation:

More especially, he [Faraday] opposed the action of forces at a distance, the assumption of two electric fluids and of two magnetic fluids, and in like manner, all hypotheses which contradicted the law of conservation of force, of which he had an early presage though he singularly misapprehended its mathematical form.

It was this criticism which Agassi thought an unscrupulous change of tune! I, on the other hand, find it most natural, in view of the interpretation suggested here. Having fixed his own concepts, Helmholtz does not realize that his own 'Kraft' had ever been in a state of change. Faraday had contributed to science important discoveries involving his concept of force—it could not have been that he did not really see clearly what forces were; thus the only thing he could find as an explanation was the much repeated criticism of Faraday's mathematical misapprehensions.

To conclude this chapter, it is fitting to add that years after his formulation of the conservation of energy in all its generality, Helmholtz did complete the circle—on the basis of his own work and by virtue of developments in thermodynamics he keenly felt the incompleteness of classical mechanics. Though he never abandoned his fundamentally mechanistic philosophy, one of his last original pieces of research was his great paper on the principle of least action, published in Crelle's *Journal* in 1887. At this stage, he was as deeply influenced by the generality and beauty of Hamilton's principle as were de Broglie, Schrödinger and Feynmann in our century. The importance of this development, and its place in Helmholtz's philosophy of physics, is well-documented in the chapter on him in Ostwald's *Grosse Männer*.

NOTES

1. 'Energy Conservation as an Example of Simultaneous Discovery', in M. Clagett (ed.), *Critical Problems in the History of Science* (Wisconsin, 1955).
2. E. C. Watson, 'Joule: Only General Exposition of the Principle of Conservation of Energy', *Amer. J. Phys.* 15 (1947), p. 383.
3. ibid.
4. ibid.
5. G. Greene, 'Applications of Mathematical Analysis to Electricity and Magnetism', in *O.E.D.*, under 'Potential'.
6. Philip P. Wiener (ed.), *Selections from Leibniz* (Scribner, 1951), p. 182.
7. This translation into English has been prepared for Dr Stephen Brush, who was kind enough to let me use the MS.

8. Per. F. Dahl, 'Ludvig Colding and the Conservation of Energy', *Centaurus* **8** (1963), 174.

9. L. Colding, 'On the History of the Principle of the Conservation of Energy', *Lond. Phil. J.* (1864).

10. ibid.

11. J. R. Mayer, 'Bemerkungen ueber die Kraefte der unbelebten Natur', *Ann. der Chem. Pharm.* (1842). Reprinted in *Raum und Zeit* Ed. E. Wildhagen (Deutsche Buch-Gemeinschaft, Berlin).

12. The translation appears in an essay called 'The Forces of Nature', in the collection edited by Youmans, *The Correlation and Conservation of Forces* (Appleton, N.Y.C., 1865).

13. See ch. I, note 11.

14. Q and q are tangential velocities; R and r are distances.

15. All quotations in English are from John Tyndall's translation, *Taylor's Scientific Memoirs* (1854).

16. As has been shown above.

17. Anyone using the expression 'logic of scientific discovery' in such a context, is rightly understood to mean that he accepts Popper's unequivocal 'no' to the question: 'Is there a logic of scientific discovery?' Concerning qualifications I refer the reader to N. R. Hanson's paper 'Is there a Logic to Scientific Discovery?' in *Current Issues in the Philosophy of Science*, Ed. Herbert Feigl and Grover, Maxwell (Holt, Rinehart and Winston, 1961); and also to the discussion and comments by P. Feyerabend in the same book. However, the sense in which I use it here, though naturally Popperian, is best expressed by E. Meyerson, again in his *Identity and Reality*:

> Our reason is competent to scrutinize everything except itself. When I reason I am really powerless to observe the action of my reason.

Naturally, not all introspection is denied but one has to go beyond one's train of thought in order to realize what has been happening. Some never reach this point, while for others it is only a question of time.

18. From 'Energy and Thermodynamics', ch. viii of Poincaré's *Science and Hypothesis*.

19. See L. Pearce Williams, *Michael Faraday* (Basic Books, 1966).

20. ibid.

21. Michael Faraday, 'On the Conservation of Force', *Proc. Roy. Soc.* (1857), republished in *Phil. Mag.* (1857). Faraday's theory on the conservation and correlation of forces underlies his charming Royal Institution lectures for 'a juvenile audience' called: *On the Various Forces of Nature* (The Viking Press, N.Y., 1960).

22. Pearce Williams, *Michael Faraday*, p. 511.

23. ibid.

24. ibid., p. 514.

25. Helmholtz, *Wissenschaftliche Abhandlungen* (2 vols. Leipzig, 1881).

26. ibid.

27. Does he want to bring to mind the Hobbesian conatus and, above all, the Newtonian associations?

28. Planck, *Das Prinzip der Erhaltung der Energie* (1913).

29. Charles Brooke, 'Force and Energy', *Nature* (1872), p. 122.

30. John Moore, 'The Conservation of Energy not a Fact, but a Heresy of Science', *Nature* (1872), p. 180. This is Moore's reply to Brooke's article. His first article appeared under the title 'The Heresies of Science' in the *London Quarterly Review* (1872).

31. Émile Meyerson, *Identity and Reality*, Dover. On p. 196, he says, talking of the *vis viva*:

'The declarations were clearly understood in their general sense, as can be ascertained by the manner of thinking in the eighteenth century, on the part of the men who were not physicists by profession, such as Haller and Voltaire.'

32. J. Nicolson, 'The Conservation of Force', *Nature* (1871), p. 47.

33. H. v. Helmholtz, Preface to the German ed. of W. Thomson and P. G. Tait's *Natural Philosophy*, trans. Professor Crum Brown and published in *Nature* (1874), p. 149.

34. H. v. Helmholtz, Preface to the German ed. of J. Tyndall's *Faraday as a Discoverer*, published in *Nature* (1870), p. 51. See also Helmholtz's 'Faraday Lecture', held before the Chemical Society meeting at the Royal Institution, and published in *Nature* (1881), p. 535.

VI
THE INSTITUTIONAL SETTING

According to the schema for the growth of knowledge as outlined in the Introduction, Chapters II–V deal mainly with the necessary conditions for the emergence of the concept of energy, that is mainly with the body of scientific knowledge. In the next chapter I shall deal with the hard core of Helmholtz's research programme, namely the development of his scientific metaphysics, its origins and growth. Scientific metaphysics is the core of a research programme;[1] it belongs to the body of knowledge; however, it is so interwoven with the image of science that it is difficult to deal with them separately. What remains to be done—and here unfortunately only in an outline form—is to pose open questions rather than to answer them: what were the purely external influences which supplied the sufficient conditions for this huge step in the growth of knowledge? In other words, what was the institutional setting for this activity? Then, what was the image of science, namely what was *thought* of as a legitimate scientific activity which made it possible for such questions to be asked?

Let us first approach the institutional problem. Since the appearance of Joseph Ben-David's book *The Scientist's Rôle in Society*, we have at last a definitive book on this problem, so I shall give here a mere summary.

Europe

Any division of the past into periods known under various names is arbitrary, and depends mostly on the point of view taken. Yet there is some broad division which, for the last three hundred years has been widely accepted, and does not seem much to be debated nowadays. On this division modern science begins roughly with Copernicus (1500) and reaches its first stage of synthesis with Newton. Newton will serve as a borderline figure whatever our point of view may be, and whether we put him as 'the last of the magicians'[2] or the first mathematical physicist. By this same technique of rough division, we have to connect inseparably the nineteenth century with ours. The problems we try to solve in physical science have mostly been asked in the nineteenth century. Our century is often called the century of biology (this word was first used in 1802);[3] the idea of progress became a commonplace among laymen and ceased being an abstract philosophical tenet only in the nineteenth century. Some call our century the 'century of psychology', for the presuppositions underlying modern psychology were hammered out in the nineteenth century; indeed the very phrase which is bedevilling psychoanalysis, and supplies fuel to its critics, 'mental energy', was created by that fascinating science called psychophysics. This field was an interesting outgrowth of the combined theories of 'energetics' and the Naturphilosophie's conception of 'force'. Every one of the above disciplines seems to me to be not less epoch-making in the long run than the typical twentieth-century contributions like relativity theory, quantum mechanics, nuclear fission and fusion, or molecular biology. Perhaps some trends in sociology form an exception.[4] In other words, I do not think that if a nineteenth-century scientist met a twentieth-century one for a long scientific exchange of ideas, there would be any difficulty of communication: the basic presupposition to both would be that Nature is in principle understandable, that most phenomena will sooner or later be explained in terms of physics (in the broadest sense), and if one of them happened to be religious, they would clearly draw a line of demarcation between his science and his religion.

In order to give a comprehensive picture of the intellectual background against which we must view the emergence of the concept of energy, the best and easiest would be to rewrite from a 1870s point of view all the four volumes of Merz's *A History of European Thought in the Nineteenth Century*. Instead, I will simply presuppose this and Ben-David's book and emphasize only a few most important and new aspects of this background.

The reason that we deal only with England, France and Germany, is that except for a few individual contributions, the bulk of the scientific activity took place in these three countries. Italy, whose contributions had been so numerous and brilliant in the seventeenth century, was torn in civil strife, and after the Napoleonic conquests, those few who did take up science migrated slowly towards Paris. There were no great European universities with an emphasis on science; the few names that come to our minds of scientists who were not from England, France or Germany, are Mendeleeff Mendel, Colding, Bolyai and Lobatschevsky. Some of the great academies of science and other institutions which had been founded in other countries (like the St Petersburg Academy, or that in Vienna) had declined very much in the nineteenth century and not much progress in science took place in them. More important than all these characteristics of the nineteenth century, there is the fruitful interaction between science and philosophy which has already been pointed out in the introduction. Though this interaction began to flourish in the eighteenth century, it became a powerful factor only at the turn of it, for as we shall see, in these three countries was prepared the ground for the emergence of the concept of energy.

France

Paris, in the first half of the nineteenth century, was the scientific capital of the world, and the seat of the government was the École Polytéchnique. Here the greatest mathematicians of the age studied, taught and created. This was the country where chemistry had been reformulated by Lavoisier a decade before, where Newtonian science had been brought to the zenith by Lagrange and Laplace, where the new mathematical foundations of a theory of heat were laid down by Fourier, Cauchy and Sadi Carnot. There are historical-sociological explanations for this development, and also for the fact that theoretical physics almost disappeared after 1820.[5] All these only emphasize the difference between the character of science in France and that in Germany: French science was concentrated in one place, in Paris, exercised by individuals who were creative in one special field, founded no 'school' and perhaps with the sole exception of Auguste Comte, did not have a philosophical system.

Ever since the time of Colbert, the French have understood that science should be government-supported and serve the nation as a whole. This rather utilitarian attitude set the tone for French science for the next two centuries. Already in the days of the Académie des Sciences[6] and the early Royal Society, one could see the different attitudes developing. The Royal Society dealt with practical problems, and applied physics; Newton never really represented the spirit of the Royal Society. Its members were mainly well-to-do upper-middle class citizens, university professors and amateurs. The Académie des Sciences from the day of its foundation was a much more aristocratic institution, and accordingly much more devoted to the interests of its kingly sponsor. In so far as it did undertake practical assignments like the survey of the kingdom which then led to the measurements of the length of the seconds pendulum and the variation of the gravity at different places on earth, these were government sponsored activities, with special interest taken in them by the king or one of his ministers. These studies again led to many mathematical studies which were supported in order to give theoretical explanations of the results brought home by the various expeditions. Thus came Huygens's study on the pendulum, and the various studies on the figure of the earth (especially Clairault's). The same trend continues in the eighteenth century, when an additional element has to be taken into account: the most abstract mathematics became fashionable and the keenest minds concentrated on it. The various problems announced by the Académie des Sciences as prize essays resulted in the most important mathematical and physical works of the century. And besides this, the great representatives of the natural philosophy were the same men as those who gave the tone in the literary circles. Merz says:

Whatever eminence German science may have gained in this century from a purely literary point of view, through the works of A. von Humboldt, or English science through those of Darwin, the history of both literatures during the eighteenth century can be written almost without any reference to science at all—so small the direct influence of such giants as Newton and Leibniz on the popular mind. But who could exclude from a history of the elegant literature of France the names of Voltaire, of Buffon, of d'Alembert, or of Condorcet?[7]

Already before the Revolution there were many great schools in France, with nothing comparable found in England or Germany. Just to mention a few—the College of France, where Gassendi and Roberval taught, in the seventeenth century, and Lalande and Daubenton in the eighteenth; the Collège et École de Chirurgie, the Jardin des Plantes (here were Buffon, Lemonnier, Daubenton and Fourcroy), the École Royale des Mines and many, many others. To centralize education and science even more, the Paris academies had also their representatives and connections in the provinces. Independent academies of science (Bordeaux, Montpellier, Toulouse, Beziers) were affiliated with the Académie des Sciences. At that time began the later famous system when every professor's dream and course was to gravitate slowly towards Paris.[8]

It was this system on which the later German hierarchy of universities became modelled, with the important political exception, that until the late nineteenth-century the German universities belonged to independent little duchies and states.

The Revolution again did much in its turn to promote institutionalized science, and the establishment of the École Polytéchnique became the most important step in the history of nineteenth-century French science. The role of Bonaparte is also very significant in these developments. Both Bonaparte's influence, and the concentration of French science around its great institutions is excellently covered by Maurice Crosland's *The Society of Arcueil*. Again there is no reason to give a summary of it here.

The École Polytéchnique is interesting from another point of view. The most characteristic feature of French science in the nineteenth century was that it was mainly mathematical. Two kinds of mathematics poured forth from the élèves of the École Polytéchnique: applied mathematics (or as we may wish to call it, theoretical physics) and pure mathematics. The theoretical physics, with the notable exception of the work of Ampère, was mainly the synthesizing sort, not what Herivel called 'creative science' or what I would prefer calling 'concept-creating' science. On the other hand a great part of the pure mathematics, mainly the beginnings of the various branches of abstract algebra, were of the most creative type. For this second phenomenon I do not offer even a tentative explanation. Too little is known, and even less is known to me on the workings of the creative mathematical mind. But for the first kind it is clear that very few new physical concepts grew out of the activity of the mathematicians of the École Polytéchnique. And an interesting coincidence has to be noted here—the only philosophical school which was created in France in those days was that of positivism, with Comte, its creator, himself a student and a teacher of the École Polytéchnique. To claim any sort of causal connection here would need further study.

England

Bulwer-Lytton characterized the nineteenth century in his *England and the English* by saying: 'Every age may be called an age of transition—the passing on, as it were, from one state to another never ceases; but in our age the transition is *visible*.' But this visible, rapid transition was greatly different in the three countries. In England 'the sacred thirst for science', as the *Edinburgh Review* called it in 1825, and the rising importance of science, coincided with the gradual democratization of English political life, and with the growing conviction that 'the scientific glory of a country may be considered, in some measure as the indication of its innate strength' (H. Davy). In this spirit, English science (if the isolated, individual contributions are not considered) was concentrated in the scientific societies which had a strong bent towards the practical side of science, and around the new industrial centres. Indeed, one of the shapers of nineteenth century England, Benjamin Disraeli, wrote in *Coningsby* in 1844: 'It is the philosopher alone who can conceive the grandeur of Manchester and the immensity of the future'.

These quotations thrust us directly into the middle of the perennial problems of interaction between science and technology, the influence of the Industrial Revolution on science. Or as someone has put it 'the steam engine has done more for science than science has done for the steam engine'. There are many other related topics. I shall not attempt to sum up the various historical approaches on this subject. Bernal's *The Social Function of Science* can serve almost as a textbook of the one approach, while all those great intellectual historians Brunschvicg, Meyerson, Cassirer and Koyré represent the other. That there is a mutual influence and great cross-fertilization between science and technology is not very much doubted: the question is where does one fix that influence. As a tentative and superficial answer, I would say that technological advances, and inductively arrived at conclusions serve as necessary conclusions, but by no means sufficient ones. We shall deal later with the claim that the realization of the impossibility of a perpetual motion is the father of the conservation of energy principle. My argument there will be that it is implied by the conservation of energy, but there was no immediate connection between the establishment of the principle and the acceptance of that impossibility. In the same vein I would claim that though the steam engine supplied the necessary concept to Sadi Carnot, it had no direct influence on his conception of reversibility and the general directional behaviour of natural processes. No doubt both Davy and Faraday learned very much from their visits in the mines but this had a profoundly different influence on these two. While in the case of Davy one could say that not only the Davy lamp, but even some of his electrochemical discoveries were triggered off by his technological knowledge; in Faraday's case one can say only that this enlarged the number of known facts to which he applied his preconceived idea of the conservation and conversion of all forces in nature. All this does not mean that in a roundabout way, extra-scientific developments do not have crucial influence on the development of science. To mention only one, it is all too natural to agree with Professor Paul Bairoch that one of the

two economic events which influenced the life of humanity more strongly than any other phenomena . . . was the industrial revolution, which freed societies almost completely from agricultural contingencies. . . . The industrial revolution was, in fact, first and foremost an agricultural revolution.[9]

To stress the interaction, let us remember only the impact and fame of Davy's Lectures on Agriculture and the part that agricultural questions played in Liebig's researches. But naturally, the basic problem of vital forces, which underlies the physiological thought of those times, had nothing to do with this. There is another level of interaction which is important in order to understand the pattern of English science in the early nineteenth century. In the same paper Bairoch says:

The arrival of industrial products—which are constantly being perfected at the same time as their prices decrease—has led to the disappearance of the local artisans.

These were the people who now, fired by the slogan that 'knowledge is power' wanted more and more popular education, and found it in the numerous lectures and lecture courses sponsored by the various scientific societies, above all by the Royal Institution.

What typefied the atmosphere of the Royal Institution was the attempt to bring education to the masses, and the belief that science should not and could not ever be pure science; in other words science was inseparably connected with technology, or rather what at the time was called the 'arts'. The degrees of this tradition varied: Bacon gave the philosophical authority to it, and the eighteenth-century interpretation of Newtonian philosophy was very much in the same line; Davy actually changed the

L

face of the Royal Institution from an adult-education centre for the workers, as planned and instituted by Rumford, into a popular science circle for the upper-middle class. Yet he was also a staunch believer in this philosophy and among the comments in his notebooks we find:

It is needless for us again to say that in *science* and the *arts* there is a dependence which is the source of their progression and importance. In a well organized country, power is always compound: Archimedes could not have made machines which terrified the Roman soldiers without the assistance of good carpenters and good workers in metal.[10]

It is interesting to note that this fragment, dated around 1810, shows that Davy used the word 'science' in the modern connotation some twenty years earlier than according to Merz,[11] who thinks that the word acquired its current meaning around the time of the formation of the British Association in 1831. In the same connection, one should note the difference between the English 'natural philosopher', the German 'Naturforscher' and the French 'savant'. There is a highly enlightening discussion of this again in Merz.[12] That the non-translatability of these terms is not the fault of our hindsight is clearly seen from the many examples where they are indeed left untranslated. Davy, in his short character sketches on various scientists[13] talks of Chaptal as the French 'savant'. Also his brother, Dr John Davy, frequently uses the French word to describe those French scientists with whom Davy came into scientific contact in France.

The above described scientific atmosphere does not apply only to the Royal Institution. There is one passage in Count Rumford's oft-quoted 'Experimental Inquiry Concerning the Source of the Heat which is Excited by Friction' read before the Royal Society on 25 January 1798, which, in my opinion, characterizes the Royal Society at this time and also the difference between the status of science in England as against France at the turn of the century. Having described the famous cannon-experiment he remarks:

For fear I should be suspected of prodigality in the prosecution of my philosophical researches, I think it necessary to inform the Society that the cannon I made use of in this experiment was not sacrificed to it.[14]

At the same time when France was establishing its most important scientific institutions with government support, the Royal Society still had to justify its activities by its immediate usefulness and absence of 'prodigality'.

There is no need to get into more detail on these institutions. Since Bence Jones's excellent story of the Royal Institution,[15] many studies of the scientific societies in England have appeared.[16] What interests us here more than anything is the system of higher education in England at the turn of the century, the important thing is that as far as science is con-

cerned, these institutions filled a more important place than the universities. Very typical of the English frame of mind with respect to education,
is a remark of John Davy on his brother:

His self-education, the most important part of the education of every original
mind, may be considered as commencing, not so much on his leaving school, as in
the year following, the year of his apprenticeship.[17]

I do not think that any student of the École Polytéchnique, or a German
university student in that time, would have said such a thing. But for the
England of 1800 this was certainly true. Nowhere else do we find such a
great number of self-educated original thinkers. There are branches of
science, and stages in the development of a science, when such an approach
is beneficial; on the other hand some other developments cannot take
place under such conditions. For one, mathematicians are very rarely self-
taught; mathematicians generally cluster in schools; mathematics is the
only field where one can speak of the French 'schools' while at the same
time they had no other organized bodies of opinion. The Germans had
the 'schools' in everything, while the English did not have them in any
field. Not that there were no mathematicians in England. There were the
Cambridge mathematicians, mostly doing what we would call now pure
mathematics. After the turn of the century a great number of Scottish
mathematicians also enriched that science. But these mathematicians did
not work in either of the two traditions,[18] on problems which were
dictated by physical phenomena. A notable exception is William Rowan
Hamilton, who in the field of pure and applied mathematics created a new
approach and a new language without belonging to any school or scientific centre. His work could have been a proud product of any of the great
French mathematicians, but his isolation and the complete novelty of the
conceptual framework in which he created is very typical of the lonely
English giants of the eighteenth century. There is much work yet to be
done until a real understanding will emerge for the originality of this
unique man; not only does his mathematical work need a revaluation—
it is almost forgotten largely because it has not successfully been used much
in physics—but also the philosophical framework of his thought. The
establishment of the analogy between wave optics, geometry, and dynamics is a kind of 'creative science'[19] which does not grow out of
mathematical reasoning, or experimental evidence, but the highest level
of philosophically motivated conceptual thinking.

Those scientists who did work in the framework of the Newtonian
tradition did not exhibit much mathematical talent. Few tried to compete
with the French mathematical work on the completion of the programme
of the *Principia*, and those who took up corpuscular philosophy had not
yet reached the stage of their science where much mathematics would

have been needed. Molecular physics and chemistry were in the first stages of their development and their task was to establish the most elementary relations and to give crude first estimates of some atomic parameters. In addition, it is true, most of these did not have much appreciation for the mathematical mode of thought, and were hardly aware of what could be done with mathematics. All this naturally changed around 1850.

There is at least one exception to the chemists who showed no understanding for mathematics (especially the English ones). Davy in two places refers to the mathematical future of chemistry, which indicates very clearly his broadmindedness and clear vision of the development of science. Describing a meeting with Laplace in France, in 1813, he says:

On my speaking to him of the atomic theory in chemistry, and expressing my belief that the science would ultimately be referred to mathematical laws, similar to those which he has so profoundly and successfully established with respect to the mechanical properties of matter, he treated my idea in a tone bordering on contempt, as if angry that any results in chemistry could, even in their future possibilities, be compared with his own labours.[20]

My point was illustrated by the first clauses of the above passage. The only reason I quoted more from it was to underline again the intellectual gulf between the corpuscular philosophers, of whom Laplace was one of the greatest, and who continued the work of Newton and brought it to new heights, and those who, believing in one or the other of the physical theories of heat, dealt with problems of chemistry. At this time, chemistry was still suffering from its undistinguished ancestry, namely, alchemy, and from the bad reputation of those who dealt in it, and developed vague, experimentally unjustified theories.

One could now easily quote document after document on the *Decline of the State of Science in England* (as Charles Babbage called his book constituting a violent attack on the Royal Society, published in 1830) and as against this on the distinguished line of English geniuses. But all this is not to the point here. What is important for us is that England did not offer any centre of higher education where an inquisitive mind would sooner or later run into all—and the emphasis is on all—open problems in the various sciences and be encouraged to see a connection between them. There were no scientific laboratories where the most brilliant young people met and joined their forces on a common enterprise, led by much debated and philosophically motivated leaders of science. In other words, England gave to the world such great ideas which by their character could come only from original individual thinkers. In all probability, had Faraday been born in Germany or France, had he ever come into any contact with science, he would have very early been stifled, his mathematical inability regarded as a debilitating *manco*, his metaphysical ideas ridiculed, his manuscripts unprinted. But this is idle speculation. The combined mathematical and physical knowledge, the awareness of problems in the animate and inanimate world, and above all the philosophical motivation, all of which were common to a great many German scientists, were not to be attained in England in the early nineteenth century.

Germany

The law of conservation of energy could not be born either within the institutional framework of France or that of England. The situation was very different in Germany. This was the country where the great philosophical systems of Kant, Fichte, Schlegel, Hegel and Lotze were conceived, at a time where in the other two countries nobody tried to construct a complete philosophical system. But even more important than that, a vast network of universities was founded, with a rigid hierarchy where philosophical faculties were established, and where the philosophers expounded their systems. The broad foundation in these universities was such that whatever the student studied he could not have avoided facing sooner or later the great metaphysical problems posed by the various 'Weltanschauungen'; while in England 'science' was pursued, or in the traditional spirit 'natural philosophy' was taught, German 'Philosophie' covered the whole of the human intellectual enterprise. Speculation was encouraged; even today, after the great, late nineteenth-century battle to erase the last remnants of the influence of the 'Naturphilosophie' (or rather of a degenerated, ridiculously trimmed version of it) 'speculation' does not cause such contempt in Germany as it does in England. It was not only the rigid, much ridiculed German 'Geheimrat' system, which made people speak of the 'school' of one or the other of the great philosophers or scientists. It was rather the fact that indeed these 'schools' or 'laboratories' represented a complete philosophical system, and every student had to take a stand towards them. One could not have worked in Gustav Magnus's physical laboratory, or in Liebig's laboratory, without taking a deeply considered philosophical approach to Kant's epistemology or to the question of the mechanistic-vitalistic controversy. In this atmosphere one could very well not separate experimental data from highly speculative hypotheses.

Far be it from me to claim that this in itself is good or bad. What I do claim is that exactly the atmosphere which was detrimental to many other scientific problems (like for example for the first formulations of electrodynamics: indeed most German theories failed with the partial exceptions of Neumann, Fechner, Weber, while the English, French and the isolated Ørsted did the work) was indispensable for the establishment of the conservation of energy principle. i.e. the emergence of the concept of energy. In these universities the 'schools' of Liebig, Wöhler, Johannes Müller, Weber and Gustav Magnus were founded. And here it was that the further prerequisite—namely, the cross-fertilization between physical and physiological thought—was made possible. Moreover, it was here that biology was born:

in the hands of German students chemistry and physics, botany and zoology, psychology and metaphysics, have laboured from different and unconnected beginnings to produce that central science which attacks the great problem of organic life, of individuation, and which states the immediate conditions of consciousness.[21]

In these universities matured men like Schwann, Brücke, Dubois-Reymond and Helmholtz. Their shared background consisted of a vast reading in the philosophers and a readiness to face their questions, an awareness of a deep-seated connection between all the sciences and a tendency to seek for coherence between their philosophies and their scientific knowledge. Helmholtz's father was a close friend of the younger Fichte, himself a professor of philosophy; Helmholtz himself, according to his own testimony, was deeply influenced by both Kant and Fichte. That, in addition to this, Helmholtz's education in physics and his mathematical ability made him the ideal man for the task awaiting him, has been shown above.

I do not think that further commentary on the German scientific atmosphere is necessary. More has been mentioned when dealing with the various aspects of Helmholtz's career. The scientific spirit in the laboratories of Johannes Müller and Gustav Magnus taught us more about Germany than any description here could possibly give.

NOTES

1. For this terminology, see I. Lakatos, 'History of Science and Its Rational Reconstructions'; and my critical reply, 'Boltzmann's Scientific Research Programme and Its Alternatives' in Y. Elkana (ed.), *The Interaction between Science and Philosophy* (Humanities Press, 1974).
2. An expression of J. Maynard Keynes in his article on Newton.
3. The term 'biology' was first coined in 1802 by G. R. Treviranus according to Merz, *History of European Thought*, vol. i, p. 193.
4. But even here it is questionable whether they could be traced back to the nineteenth century.
5. J. Herivel, 'Prerequisites for Creativity in Theoretical Physics', *Scientia* 60 (1966).
6. See Roger Hahn's impressive *The Anatomy of a Scientific Institution, The Paris Academy of Sciences, 1666–1803* (California, 1971).
7. Merz, *History of European Thought*, vol. i, p. 105.
8. ibid.

9. Paul Bairoch, 'Original Characteristics and Consequences of the Industrial Revolution', *Diogenes* 54, p. 47.

10. John Davy, *Fragmentary Remains*, p. 116.

11. Davy, *Collected Works*, vol. i, p. 89.

12. ibid.

13. ibid., vol. i, p. 168.

14. Count Rumford, 'Experimental Inquiry Concerning the Source of the Heat which is Excited by Friction', *Phil. Trans.* lxxxvii (1798), p. 80. Also reprinted in several other journals simultaneously.

15. Bence Jones, *The Royal Institution* (London, 1971).

16. T. Martin, *The Royal Institution* (London, 1941) and P. Schofield, *The Lunar Society of Birmingham* (Oxford, 1953).

17. Davy, *Collected Works*.

18. i.e. the mechanical tradition growing on the *Principia* and developing into rational mechanics, and the theory of matter taking up the suggestions of the queries at the end of the optics and developing into chemistry and molecular physics. See below.

19. Using the terminology of Professor J. Herivel in his 'Prerequisites for Creativity in Theoretical Physics', *Scientia* 6 (1966). This is a brave, though not completely successful attempt to define creative science.

20. Davy, *Collected Works*, vol. i, p. 168.

21. ibid., p. 95.

VII

THE PHILOSOPHICAL
BACKGROUND

Helmholtz's scientific temperament

The story of the principle of conservation of energy has been told many times. The most fruitful years in this development, 1820–50, have been examined from various points of view;[1] but whatever the point of view may be, there is unanimous agreement that the first mathematical formulation of the principle in all its generality is that of Hermann von Helmholtz in 1847. We have seen that what Helmholtz actually did was to formulate clearly the conservation of mechanical energy and then to show that all the various 'forces of nature' can be reduced to that form, that is, subsumed to the conservation principle. He thus created the very general concept of energy as the one entity that is being conserved under all circumstances in a fundamentally mechanical world. Moreover, the generality of Helmholtz's treatment is underlined by the fact that, to the best of my knowledge, for the first time the principle is treated not only as a law of nature covering all phenomena, but, to use an expression of Maxwell's, as 'science-producing doctrine':[2]

The object of the present memoir is to carry the same principle in the same manner through all branches of physics; partly for the purpose of showing its applicability in all those cases where the laws of the phenomena have been sufficiently investigated, partly supported by the manifold analogies of the known cases, to draw further conclusions regarding laws which are as yet but imperfectly known, and thus to indicate the course which the experimenter must pursue.[3]

It is time now to look closer at my claim that Helmholtz belonged to those physicists whose scientific temper involved commitment to *a priori* beliefs, and parallel to that, to trace the philosophical influences on the development of his ideas about conservation of force. There is no direct evidence at my disposal which could carry my point. There are no sweeping Helmholtzian statements about his metaphysical commitments, or about his pre-scientific beliefs. On the contrary, Helmholtz is so much a typical nineteenth-century scientist, that he tries to eradicate the last remnants of German Naturphilosophie, which in his opinion, and in the opinion of his contemporaries, did enormous harm to the development of German science. Along with this goes a loud empiricism, a periodically repeated denouncement of hypotheses and conjectures. And yet, all this can be found in Helmholtz's own writings.

In a previous chapter, I have demonstrated the inherent vagueness in the verbal formulation of the conservation law—'Die Erhaltung der Kraft'. If we recall all the various meanings attached to the word 'Kraft', we have to admit that the formulation, if not prescientific, was at least nonscientific. Do we have to take Helmholtz at his word, accepting that the whole idea had its basis in an inductive realization that one could not construct a *perpetuum mobile*? Helmholtz writes:

So stiess ich auf die Frage: 'Welche Beziehungen muessen zwischen den verschie-denartigen Natur-Kraeften bestehen, wenn allgemeinh in kein *Perpetuum Mobile* moeglich sein soll?'

In this way I came to the question: 'What must be the relations between the various forces of nature if no *perpetuum mobile* should be possible?'[4]

And in his 1881 notes to the 1847 paper he adds:

Uebrigens ist dieses Gesetz, wie alle Kenntnis von Vorgaengen der wirklichen Welt, auf induktivem Wege gefunden worden. Dass man kein *Perpetuum mobile* bauen d.h. Triebkraft ohne Ende nicht ohne entsprechenden Verbrauch gewinnen koenne, war, durch viele vergebliche Versuche es zu leisten, allmaehlich gewon-nene Induction.[5]

Besides, this law, like all knowledge about proceedings in the real world, was discovered inductively. That no *perpetuum mobile* can be built, i.e., that no motive power can be gained endlessly without corresponding expenditure, had become, through many unsuccessful attempts, a gradually obtained induction.

Naturally, Helmholtz was convinced that no perpetual motion machine could be built. But so was the French Académie, as we have seen, in 1775, when it decided not to consider any further attempts to construct a perpetuum mobile. As Helmholtz formulates it, he did not look for a law which would be a generalization of this impossibility. He knew only that whatever law he was going to suggest must not contradict this impossi-bility. But in this, there is nothing new. Every well-educated physicist in the last two centuries avoided a physical theory which would imply a *perpetuum mobile*. The difficulty was naturally that until the connection between the impossibility of a *perpetuum mobile* and the general principle of the conservation of energy became clear, in some new theories there was no way at all to foresee whether or not a *perpetuum mobile* would be implied by it. Count Rumford was as sound a scientist as any of his time; he certainly knew that it would be useless to try and construct a *perpetuum mobile*, and yet his real argument against the caloric theory was that he had an inexhaustible source of heat; heat could not be matter because matter obeyed an all-embracing conservation law. The implications of his argument, as we have seen, are that as heat is not matter, it can be only a quantity which is not being conserved in nature.

Besides the impossibility of gaining work from nothing—even if it is not any less general than the principle of conservation of energy—it is, as Planck pointed out, only half the story. In his opinion, this conviction worked psychologically only in one direction. People never realized that one could not lose work infinitely, and even thought that Count Rumford's famous cannon experiment teaches us the contrary. On the other hand, for a clear understanding of the conservation of energy principle, one has to see that the impossibility works in both directions.[6]

Thus, referring back to what was said when reviewing the influence on Helmholtz of the double-tradition of mechanics, it was certainly clear to him that, whatever form the law that connects the 'forces of Nature' will acquire, it must satisfy this most self-evident and long-known condition.

The status of the concept 'Kraft': 'force' or 'energy'?

It was not the case that Helmholtz had realized, as an inductive inference, that energy had to be conserved, while others inferred wrongly that it was force (in our sense of the word) which was conserved. There were no two clearly formulated, mathematically expressible conservation laws—a law of conservation of force, and a law of conservation of energy—so that, due to the work of Helmholtz and many others, the proof of the second was found correct, while the proof of the first was not satisfactory. In fact, no proof of a conservation of force, as we understand force, can even be formulated; one could talk of a 'conservation of force' in English as long as the definition of the concept was on the level of precision of Faraday's, or in German about the 'Erhaltung der Kraft'; before the clear statement of the conservation which led to the establishment of the concepts; 'Kraft' simply became 'force' and the conserved quantity was called 'Energie'. I have already observed that had Faraday tried to formulate his theories in a mathematical language, he would in all probability have discovered the principle of conservation of energy in all its generality exactly as Helmholtz succeeded in doing, or he would have had to abandon his theory about the conservation of force. Helmholtz, having been a mathematician of the first rank, did aim at an immediate mathematical formulation which resulted in his proof, and, along with it, in a clarification of the concepts. It is sufficiently clear that a law formulated with such concepts and, lacking mathematical clarity to such an extent, could not have been drawn as an inductive inference. Certainly, one could arrive inductively at the conclusion, that a *perpetuum mobile* is impossible, or that the *vis viva* is being conserved in perfectly elastic collisions, or at the conservation of the sum of the 'tension' and 'life force', or even the mechanical equivalent of heat. But one certainly could not arrive inductively at a generalization, that 'something in Nature must be conserved', at the level where it is treated as a science-producing principle. The principle of conservation of force was not disproved; it was labelled a different verbal formulation of the true principle, and then the whole thing faded away, as generally happens with superseded scientific theories. I would like to point out, that recently more and more attention has been given to this vague state of formulation of scientific concepts. Professor Laszlo Tisza published, in the *Review of Modern Physics*, an important paper called 'The Conceptual Structure of Physics'. He does not deal with change of concepts on an individual or on a psychological level, but with the logical status of conceptual dynamics, which bears some resemblance to what I have been trying to prove. Concepts, in his opinion, undergo differentiation and integration when great scientific discoveries occur. If I understand him correctly, he would say for the case treated here that, through the establishment of the principle of conservation of energy, the concept 'Kraft' differentiated into 'force' and 'energy', while the concepts of 'puissance motrice', 'Spannkraft', 'Arbeitskraft', 'vis viva', 'vis morte', and many others, integrated into the concept of energy. Suggestively enough, Professor Tisza quotes a motto by H. A. Kramers, which I found irresistible and must repeat yet again:

My own pet notion is that in the world of human thought generally, and in physical science particularly, the most fruitful concepts are those to which it is impossible to attach a well-defined meaning.[7]

From the point of view of epistemology, this vague stage in the development of science was admirably treated by Emile Meyerson;[8] indeed, it is in his spirit that this work was undertaken.

That the concept of energy could have been defined clearly only after the law of its conservation had been established, that is, that the principle of 'Erhaltung der Kraft' must have been a vaguely formulated belief, or persuasion—something which shows that Helmholtz must have adhered to it, very much *a priori*—was clearly seen by Planck:

Der Begriff der Energie seine Bedeutung fuer die Physik erst durch das Prinzip gewinnt.[9]

The concept of energy gains its meaning for physics only by the principle [of its conservation].

The campaign against metaphysics

Helmholtz tried to convince himself and others of his thorough empiricism, which meant for him an anti-metaphysical and anti-hypothetical approach.

We have already seen that he tried to prove that the conservation principle had been arrived at inductively. Even more sharply, he has the following to say about J. R. Mayer's discovery of the mechanical equivalent of heat in his 1881 notes to his essay on conservation:

In neuester Zeit haben die Anhaenger metaphysischer Speculationen versucht das Gesetz der Erhaltung der Kraft zu einem *a priori* gueltigen zu stempeln, und feiern deshalb R. Mayer als einen Heros im Felde des reinen Gedanken. Was sie als Gipfel von Mayer's Leistungen ansehen, naemlich die metaphysisch formulierten Schein- beweise fuer die apriorische Nothwendigketi dieses Gesetzes, wird jedem an strenge wissenschaftliche Methodik gewoehnten Naturforscher gerade als die schwaechste Seite seiner Auseinandersetzungen erscheinen, und ist unverkennbar der Grund gewesen, warum Mayer's Arbeiten in naturwissenschaftlichen Kreisen so lange unbekannt geblieben sind.[10]

Lately, the followers of metaphysical speculations have tried to stamp the law of the conservation of energy as valid *a priori*, and they are celebrating R. Mayer as a hero in the field of pure thought. What these people consider as the zenith of Mayer's contribution, namely, the metaphysically formulated would-be proofs for the *a priori* necessity of this law, will be considered the weakest point of his in the eyes of anyone used to the severe scientific method of the natural scientist. This was doubtless the reason for his work having remained unknown for so long among natural scientists.[11]

Helmholtz is carried away by his own conscious battle against the meta- physical speculator. But he, Helmholtz, had known the value of Mayer's work for a few decades when writing these lines; why did he not do any- thing to propagate it? Again, I do not think that the answer lies in petty priority resentments; Helmholtz never attacked Joule or any of the other numerous discoverers of the law. His only severe criticisms are against R. Mayer, and the one mentioned above against Faraday; both succeeded in becoming great scientists, though not on the lines of Helmholtz's avowed empiricist principles. Had he forgotten that these same argu- ments led the editor of the *Poggendorffsche Annalen* to reject his own paper, back in 1847? Let us compare what Helmholtz had to say about the status of the law of conservation with what other great physicists maintain:

One proposition only shall be alluded to as having been by some writers rather overstrained, viz., that the amount of energy in the world is unchangeable, the sum of the actual or kinetic and potential energies being a constant quantity. This may be taken as a postulate and is probably true but it is a proposition that is equally incapable of proof or disproof.[12]

This was Charles Brooke, in 1872! In Meyerson's *Identity and Reality*, we find:

What really would be a valid experimental demonstration of the conservation of energy? We should need a considerable series of experiments showing that through all kinds of change, under the most varied conditions, different forms of energy transform themselves into one another according to equivalents remaining constant within the limits of error of measuring instruments. It seems to be a demonstration of this kind that Helmholtz was thinking of in 1847 when after having furnished the double deduction . . . ended by 'the complete confirmation of the law must be considered one of the principal tasks which physics will have to accomplish in the years to come'. Even today the task cannot be accomplished. It is not even very certain that if we were able to measure very exactly all the energy known to us, as present in any phenomenon, we should find it really constant, and this for the quite simple reason that we are in no way sure of knowing all the forms of energy.[13]

So far, we have been trying to show that Helmholtz could not have been practising what he was preaching. But there is also positive evidence, that when he is not speaking as a methodologist of empirical science, he expresses himself quite differently. In the original essay, Helmholtz has given the following programme for the whole of natural science:

Das endliche Ziel der theoretischen Naturwissenschaften ist also die letzten unveraenderlichen Ursachen der Vorgaenge in der Nature aufzufinden.[14]

The final aim of theoretical natural science is therefore to discover the ultimate and unchangeable causes of natural phenomena.

Or again, a few paragraphs later:

Wir haben oben gesehen, dass die Naturerscheinungen auf unveraenderliche letzte Ursachen zurueckgefuehrt werden sollen; diese Forderung gestaltet sich also so, dass für letzte Ursachen der Zeit nach unveraenderliche Kraefte gefunden werden sollen.[15]

We have seen above that the problem before us is to refer the phenomena of nature to unchangeable final causes. This requirement may now be expressed by saying that for final causes unchangeable forces must be found.

Having thus formulated the task of physical science, and going beyond the old law of conservation of *vis viva*, Helmholtz continued to investigate how general such a limited principle could become. He goes into all other branches of physics and finally applies it to the energy of life in living organisms. Both the formulation of the programme and the way Helmholtz implemented the generalization seem to me typical of a very general, very much *a priori* principle. In this I am not alone. Réné Dugas writes about Helmholtz:

In 1847 Helmholtz published in Berlin a paper on the Conservation of Force. From the philosophical, not to say the metaphysical point of view, Helmholtz assigned to the theoretical sciences the task of inquiring into the 'constant causes' of phenomena. . . . In some things Helmholtz appears a Cartesian. Having endowed matter with extension, and with quantity of mass, he was of the opinion that matter can only recognize changes of position in space, that is movement. But matter also shows *activity* in movement. Therefore the energetic thesis is concerned with properties of 'active matter'. . . .

But it is well to remember his intention 'to reduce all natural phenomena to invariable forces. . . .' There is some illusion in the pursuit of such an ideal, and this, to Helmholtz himself, had, at least in part, the character of a wish.[16]

We shall see in the next section that what Dugas here calls Cartesian has been introduced into Helmholtz's philosophy through Kant. Nevertheless, I hope that what I have described as Cartesian has also clearly been seen to have been attributed to Helmholtz in unmistakable terms. Helmholtz, as we have seen earlier, was much occupied by Faraday's achievements. In 1870 Helmholtz wrote an essay about Faraday (as mentioned above, in the form of an introduction to a German edition of Tyndall's *Faraday as a Discoverer*), where he says:

Matter, Force, Atoms and Imponderables. . . . It was these ideas that Faraday sought in his riper labours to purify from everything theoretical, which was not the true and immediate expression of the facts.

This is not exactly the case with Faraday; moreover, let us recall the warm praises Helmholtz was pouring on him in his Faraday Lecture in 1881:

. . . [Faraday] discovered induced currents; he traced them out through all the various conditions under which they ought to appear. He concluded that somewhere in a part of the space traversed by magnetic force there exists a peculiar state of tension, and that every change of this tension produces electromotive force. This unknown hypothetical state he called provisionally the electrotonic state . . . and now with quite a wonderful sagacity and intellectual precision Faraday performed in his brain the work of a great mathematician without using a single mathematical formula. He saw with his mind's eye that by these systems, . . . perhaps a Clerk Maxwell, a second man of the same power and independence of intellect, was necessary to reconstruct in the normal methods of science the great building, the plan of which Faraday had conceived in his mind.[17]

Helmholtz did not therefore change between the years 1870 and 1881 (as seen by these last two quotations); let us see one more sentence, which appears a few paragraphs before the last one:

His principal aim was to express in his new conceptions only facts, with the least possible use of hypothetical substances.[18]

I wish to make it very clear once more that I do not try to pass here any value judgement or to grade minds of the stature of Helmholtz's or Maxwell's. Both of them were intellectual giants, and being great scientists, neither was dogmatic or *a priori* in their specific physical theories. The emphasis here is on specific physical theories; for Helmholtz was certainly *a priori* in his general approach to conservation principles. On the other hand, when trying to decide what mathematical form the conserved entity was to take, his method was strictly scientific (as we understand scientific, in this case as against *a priori*, without entering the labyrinth of questions, as to what exactly a scientific procedure is). Unlike Helmholtz, Maxwell, to the best of my knowledge, did not use such vast *a priori* principles for his premises. The intellectual tradition of Helmholtz is the one I called Cartesian, or concept-creating. Again, by Cartesian is meant, not the physics of Descartes, but his fundamental belief about the general behaviour of Nature.[19]

What the above passages are supposed to show is only that Helmholtz tried to seem an extreme inductivist, to be himself like Maxwell who 'worked with the normal methods of science', as we saw Helmholtz describing it. I certainly do not consider the above passage as proof of Helmholtz's metaphysical beliefs, but only proof of how very great the discrepancy is between his professed scientific method and his actual one. The only conclusion I want to draw from this is, that we cannot take too seriously the 'Baconian' image[20] that Helmholtz built for himself.

Not only did Helmholtz formulate the law of conservation of energy, not only did he, in the last phase of scientific activity, return to basic mechanical principles by expounding the principle of the least action, but he also indulged in brilliant speculative deductions; some of these were proved wrong and were forgotten—for example, his electromagnetic theories. Some, however, proved to be astonishingly correct. In the same Faraday lecture he wrote:

Now the most startling result, perhaps of Faraday's law is this: if we accept the hypothesis that the elementary substances are composed of atoms, we cannot avoid concluding that electricity also, positive as well as negative, is divided into definite elementary portions, which behave like atoms of electricity. As long as it moves about in the electrolytic fluid each atom remains united with its electric equivalent or equivalents. At the surface of the electrodes, decomposition can take place if there is sufficient electromotive power, and then the atoms give off their electric charges and become electrically neutral.[21]

The influence of Kant

Kant's influence on Helmholtz has been acknowledged by himself openly, but I will touch on this only briefly. More detailed treatment would involve a profound analysis of the Kantian conception of causality, the connection between this and the absolute rule of conformity to law, and a review of causal philosophy against the background of nineteenth-century 'mechanical philosophy'. Kant's influence on Faraday had been shown clearly by L. Pearce Williams, and that on other scientists in the nineteenth century, by Mary B. Hesse in her *Forces and Fields*. I will confine myself to a few passages from Helmholtz and their analysis.

(1) Alle Wirkungen seien in der Natur zurückzufuehren auf anziehende und abstossende Kraefte, deren Intensitaet nur von der Entfernung der auf einander wirkenden Punkte abhaengt.

All actions in nature can ultimately be referred to attractive and repulsive forces, the intensity of which depends solely upon the distances between the points by which the forces are exerted.

(2) Der theoretische Theil derselben [der Naturwissenschaften], sucht dagegen die mechanischen Ursachen der Vorgaenge aus ihren sichtbaren Wirkungen zu finden; er sucht dieselben zu begreifen nach dem Gesetz der Causalitaet. Wir werden genoethigt zu diesem Geschaefte durch den Grundsatz, dass jede Veraenderung in der Natur eine zureichende Ursache haben muesse.

The theoretical portion [of natural science] seeks, on the contrary, to evolve the unknown causes of the processes from the visible actions which they present; it seeks to comprehend these processes according to the law of causality. We are justified, and impelled in this proceeding, by the conviction that every change in nature *must* have a sufficient cause.

(3) Qualitative Unterschiede duerfen wir der Materie an sich nicht zuschreiben denn wenn wir von verschiedenartigen Materien sprechen, so setzen wir ihre Verschi.denheit immer nur in die Verschiedenheit ihrer Wirkungen, d.h. in ihre Kraefte.

To matter, thus regarded, we must not ascribe qualitative differences, for when we speak of different kinds of matter we refer to differences of action, that is, to differences in the forces of matter.

(4) Die Begriffe von Materie und Kraft in der Anwendung auf die Natur nie getrennt werden duerfen . . . wir koennen ja die Materie den nur ihre Kraefte, nie an sich selbst, wahrnehmen.

In the application of the ideas of matter and force to nature, the two former should never be separated . . . matter is only discernible by its forces, and not by itself.

(5) Bewegungen und die aeusseren Verhaeltnisse, durch welche die Wirkung der Kraefte modifiziert wird, koennen nur noch raeumliche sein, also die Kraefte nur Bewegungskraefte, abhaengig in ihrer Wirkung nur von den raeumlichen Verhaeltnissen.

The only alteration possible to such a system is an alteration of position, that is motion; hence the forces can be only moving forces dependent in their action upon conditions of space.[22]

The passages quoted above, from Helmholtz's rather philosophical introduction to his 'On the Conservation of Force' (though he intended to eliminate all philosophy from this paper, as he says at the beginning)[23] reflect Kant's saying that: 'the idea of causality leads to the concept of action, this to the concept of force and thereby to the concept of substance.'[24] 'Substance' matter is not conceivable as such, but only through its forces and action. And even more clearly, we can see the connection between Kant and Helmholtz if we compare the following two passages. The first is from the *Critique*:

We know substance in space only through the forces which work in this space, either by drawing others to it (attraction) or by preventing penetration (repulsion and impenetrability).[25]

The other is Helmholtz's programme for all of natural science, in a slightly different formulation than the one given above, but again taken from his original 1847 essay:

Die Naturerscheinungen sollen zurueckgefuehrt werden auf Bewegungen von Materien mit unveraenderlichen Bewegungskraeften. . . . Eine Bewegungskraft welche sie gegen einander ausueben, kann deshalb auch nur Ursache zur Aenderung ihrer Entfernung sein, d.h. eine anziehende oder abstossende.

The phenomena of nature are to be referred back to motions of material particles possessing unchangeable moving forces. . . . A moving force which they exert on each other can act only so as to cause an alteration of their distances, that is, it must be either attractive or repulsive.[26]

Even before a deeper analysis of the Kantian influence, it is evident from these passages that they could have been taken, with slight modifications, straight out of Kant's *Metaphysische Anfangsgruende*.

Helmholtz has often spoken and written about his indebtedness to Kant. His important paper 'On the Conservation of Force' was written when he was only twenty-seven and is strongly Kantian. In later years he moved away from this influence or at least tried to do so. By 'tried'—I mean that in fact all through his life he remained true to his thoroughly mechanistic approach, which is again very much Kantian. He never really passed far from an intellectual commitment to the absolutely fundamental character of central forces; he only generalized his approach even more, when dealing with the principle of least action. This however is beyond the scope of the present analysis. Only one remark is necessary here, Emile Du Bois-Reymond, Helmholtz's friend and colleague, who shared this 'mechanical philosophy' traces the basic influence in this direction not directly to Kant but to Boscovich! Where Helmholtz really departed from Kantian philosophy was in his theory of perception and as a result, in his theory of knowledge. Helmholtz's lifelong researches led him to different opinions: Kant taught that space and time were transcendental forms of perception, and that these could be determined only by axioms, while Helmholtz's opinions are succinctly summarized by

Professor L. Koenigsberger in his famous Helmholtz lecture, given in Heidelberg in 1895:

[Helmholtz's] physiological-optical researches caused him to consider the source of the general perception of space, and he was very soon led to the conviction that only the appearance of space relations causes us to grant as self-evident that which, in reality, is a particular characteristic of our exterior surroundings and we therefore regard the axioms of geometry as laws given by transcendental perception.[27]

In 1881, when writing those very revealing notes to his original essay, which we have already mentioned several times, Helmholtz regretted the great influence of Kant on his original formulation. In the meantime he had become such an empirically minded anti *a priori* thinker, that he would probably have detected these elements even in the work of his youth. But this did not constitute a change of attitude; in the original paper and in later publications, he could indulge in philosophical speculations a few passages before and after having stated the empirical basis of the conservation principle, or having sermonized on the true scientific method which was so 'deplorably ignored' by Mayer or Faraday or others. At any rate, in those notes we find:

Die philosophischen Eroerterungen der Einleitung sind durch Kant's Erkenntnis-theoretische Ansichten staerker beeinflusst als ich jetzt noch als richtig anerkennen moechte. Ich habe mir erst spaeter klar gemacht, dass das Prinzip der Causalitaet in der That nichts anderes ist als die Voraussetzung der Gesetzlichkeit aller Naturer-scheinungen. Das Gesetz als objektive Macht anerkannt, nennen wir Kraft. Ursache ist seiner urspruenglichen Wortbedeutung nach das hinter dem Wechsel der Erscheinungen unveraenderlich Bleibende oder Seiende, naemlich der Stoff und das Gesetz seines Wirkens, die Kraft.[28]

The philosophical assertions in the introduction have been influenced by Kant's epistemological views to a larger extent that I would think now correct. It has only later become clear to me that the law of causality is in reality nothing but a presupposition that all the phenomena of nature are subject to law. The law, realized as an objective power, we call Force. A cause, according to the original meaning of the word, is that, which is, or remains, unchanging, behind the changes in the phenomena, namely, the substance and law of its action, i.e., Force.

Helmholtz here equates the principle of causality with an absolute rule in nature. In Meyerson's words:

Let us take up again the formula of Helmholtz, considering it for what it really is, that is an expression of the principle not of causality but of lawfulness [légalité]. . . . This will permit us to translate . . . Helmholtz's phrase into 'the supposition of the lawfulness [Gesetzlichkeit] of all phenomena of nature.[29]

But even more:

Helmholtz, although he had characterized conformity to law as a 'presupposition', in certain respects it is much more than that, it is a conviction.

Thus it is clear that Helmholtz tried to reduce causality to lawfulness and demanded as a first aim of science 'to reduce all phenomena of nature to the unchangeable forces of attraction and repulsion . . .'; asserting that 'the solubility of this problem is the condition for a complete comprehensibility of nature'. Without entering the very subtle philosophical analysis concerning differences between a principle of causality and a principle of lawfulness in nature, it is clear that Helmholtz's aim was to show their mutual dependence, or even identity. It is clear also that Helmholtz was influenced by Kant at least in his epistemology. On the other hand, did a great scientist, who was 'Cartesian' in his approach, need Kant's influence in order to become a firm believer in the absolute lawfulness of Nature? What that influence might have contributed was to cause Helmholtz to realize his own attitude. Finally, it is clear that this special form of a fundamental belief, a belief in a conservation law which covers all branches of physics (and of the living world too) is only one step from a clearly formulated belief in the law of causality. Mach, in his essay 'On the Conservation of Energy and its History', says:

After these considerations, now it will not be hard for us to discover the source from which this principle of the excluded perpetual motion arises. It is again only another form of the law of causality.[30]

All this was meant to underline the argument that Helmholtz was committed to a very general, vaguely formulated, conservation law, and that, as we have seen, in his own works and in the sources that influenced him, the conserved entity of such importance must have been that same vaguely defined 'Kraft' of which he was talking prior to his mathematical proof of its conservation, i.e. prior to the fixing of the concept of energy.

Once more a comparison between Helmholtz and Faraday comes to mind; while Helmholtz was leading the school of the empiricists and made loud denunciations of any metaphysical tendencies in science, Faraday was much more aware of his own attitude:

I have had doubts in my own mind whether the considerations I am about to advance are not rather metaphysical than physical. I am unable to define what is metaphysical in physical science.[31]

The aftermath

By the 1880s, the principle of the conservation of energy had become one of the most fundamental principles in all natural science. In physical literature and in the semi-popular press in England, the 'Conservation of Force' has faded out of memory. Maxwell summed up the situation very clearly in his preface to his *Matter and Motion* in 1877. (One wonders, in view of all that has been said, whether his summary was not a little premature; but even if it was, and it was not the correct picture in 1877, it was certainly the general picture ten or fifteen years later.) There we find:

Physical science, which up to the end of the eighteenth century had been fully occupied in forming a conception of natural phenomena as the result of forces acting between one body and another, has now fairly entered on the next stage of progress—that, in which the energy of a material system is conceived as determined by the configuration and motion of that system, and in which the ideas of configuration, motion and force are generalized to the utmost extent warranted by their physical definitions.

In Germany also, the concept of energy had become so fixed that nobody used the term 'Kraft' any more except for 'force'; in 1887 the Philosophical faculty of the University of Göttingen announced an essay prize, the purpose of which was to clarify whether the well-known principle of 'conservation of energy' is identical with Helmholtz's 'Erhaltung der Kraft'. As they put it:

Seit Thomas Young (*Lectures on Natural Philosophy*, London 1807, Lecture VIII) wird den Koerpern von vielen Physikern Energie zugeschrieben, und seit William Thomson (*Philosophical Magazine and Journal of Science*, IV, Series, London 1855) wird haeufig das *Prinzip der Erhaltung der Energie* als ein fuer Koerper gueltiges ausgesprochen, worunter das selbe Prinzip verstanden zu werden scheint, was schon frueher von Helmholtz unter dem Namen des Erhaltung der Kraft ausgesprochen war.[32]

With appropriate historic irony, one of the prizes was awarded to Max Planck, who wrote an exhaustive historical and philosophical survey on the principle, attributing to Helmholtz the clear concept of 'energy' before its real emergence through the proof of its conservation; that is, to the very man whose contributions set in motion a new philosophy of nature with an absolutely new concept of energy.

NOTES

1. This period has been treated fully by Professor Thomas Kuhn (in his article 'Energy Conservation') from the point of view of 'simultaneous discovery' seeking the various sociological, scientific, and other roots of this phenomenon. I do not wish to quote the extensive literature on this subject; my only repeated emphasis here is that I do not deal with the question of 'priorities'. I do not try to claim Helmholtz's conceptual priority in realizing the full implications and generality of the principle, he brought to gelation the concept of energy.

2. Maxwell, *Matter and Motion*.

3. Helmholtz, 'Über die Erhaltung der Kraft'.

4. ibid.

5. Helmholtz: notes added to the original 1847 paper, 'Ueber die Erhaltung der Kraft' in the 1881 edition of his *Wissenschaftliche Abhandlungen*. Here the translation is mine.

6. Planck mentions this in his *Das Prinzip der Erhaltung der Energie*. As to Rumford's work, let me state here only that I do not see a direct dependence of the conservation of energy principle on the mechanical theory of heat. The historical sequence seems not to be mechanical theory, conservation of energy principle, or second law of thermodynamics; but rather a heat theory which was mathematically formulated before the nature of heat was clarified, and which led to Carnot's principle (while it was vague, the exact nature of the conserved entities, as seen above); and after the independent discovery of the principle of conservation of energy, a separation of the foundations of thermodynamics into two fundamental laws.

7. Tisza, 'Conceptual Structure of Physics', *Rev. Mod. Phys.* **35** (1962), p. 343.

8. Émile Meyerson: *Identity and Reality*. Also relevant is his essay: 'Y a-t-il un rhythme dans le progrès intellectuel?' in his *Essais*.

9. Planck, *Das Prinzip der Erhaltung der Energie*.

10. ibid.

11. My translation.

12. Charles Brooke, 'Force and Energy', *Nature* (1872), p. 122.

13. Meyerson, *Identity and Reality*, p. 156.

14. 'Über die Erhaltung der Kraft'.

15. ibid.

16. Dugas, *History of Mechanics*.

17. Published in *Nature* (1881), p. 535.

18. ibid.

19. ibid.

20. ibid.

21. ibid.

22. Helmholtz, 'Über die Erhaltung die Kraft'.

23. In a letter to Émile Dubois-Reymond, Helmholtz says that he has eliminated from the essay all that 'smells of philosophy'. If this was what he retained, I wonder how metaphysical the original must have been!

24. Quoted by L. Pearce Williams, *Michael Faraday*.

25. I. Kant, *Critique of Pure Reason*.

26. Helmholtz, 'Über die Erhaltung der Kraft'.

27. Koenigsberger, 'The Investigation of Hermann von Helmholtz on the Fundamental Principles of Mathematica and Mechanics', *Ann. Rep. Smiths. Inst.* (1890).

28. Notes to the 1847 paper in the 1881 edition of *Wissenschaftliche Abhandlungen*.

29. Meyerson, *Identity and Reality*.

30. Mach, *History and Root of the Principle of Conservation of Energy*.

31. Though not undertaken here, I want to stress the point that it is not enough to trace the influences on Helmholtz back to Kant. Helmholtz read great parts of Euler's works, he read Fichte, and he read, very carefully, Goethe. Besides, as I will show elsewhere, the physical foundations of Kantian philosophy are to be found in Euler, and most of the very vague and metaphysical formulations in Kant on the concept of force derive originally from Euler. Thus the circle of philosophical influences is closed. Both Newton and Leibniz did a great part of their work in opposition to Descartes, who is their common source of inspiration. The widely differing Newtonian and

Leibnizian influences converge in the philosophical writings of Euler which exerted a decisive influence on Kant. Kant is the spiritual fore-father of German Naturphilosophie, which again, together with direct Kantian influence, contributed to the final formulation of the conservation law, and to the various theories of force, in the nineteenth century. Of these theories of force, again two widely divergent conceptual frameworks developed: that of energy and that of the field.

32. Planck, *Das Prinzip der Erhaltung der Energie.*

VIII

CRITICISM AND OPEN QUESTIONS

For the sake of a photographic picture we stopped time and looked at the intellectual environment, the body of knowledge and the image of knowledge in the 1840s in Helmholtz's Germany. Then, we saw chronologically how the interaction between the changing image of knowledge and the various scientific research programmes took place. We looked at the process of the growth of knowledge through the lens of one important case history. The greatest emphasis in the book was on the various currents in the body of knowledge which were indispensable for the establishment of the conservation principle. To some extent we followed the influence of the intellectual environment on the hard-core scientific metaphysics of the scientific research programme of Helmholtz. Also to some extent we saw how the competing images of science influenced the choice of problems of those working on them—especially how Helmholtz's commitment to the view that Nature was comprehensible, that great principles of general validity were acting in Nature and that the task of science is to understand these principles dictated his problem-choice—the principle of conservation of energy. Because of the emphasis the book is still a chapter in the history of ideas and not in comparative historical sociology of scientific knowledge. This in my opinion is a great drawback: it is only one half of the task accomplished. As a result we are still facing problems of fundamental importance to which no answer has been sought here.

The main unanswered questions are the following. What was there in the intellectual environment of Europe in general, and in Germany especially, that influenced scientific metaphysics to centre around conservation ideas? Why was it in the 1840s that the image of knowledge of the well-equipped professionals shifted so that they could accept such scientific metaphysics? Can one draw up any causal relationship between the economo-politico-social factors and the change in the image of knowledge? What was the influence of the academic institutions in these developments? All these questions have to await the further researches of more than one person. Yet one thing is clear: case studies have to be measured against theories of growth of knowledge if we wish to gain any illumination from them. And here, beyond doubt, we are out not for simplicity but for ever growing complexity.

The conservation of energy: a case of simultaneous discovery?

(First appeared in *Archives Internationales d'Histoire des Sciences* (1970), no. 90–1, pp. 31–60)

The central concepts in the physics of Newton were space, time, mass and force. By the end of the nineteenth century the central concepts in physics were space, time, mass and energy.[1]

As evidenced by Planck and Poincaré,[2] the general concept of energy became meaningful only through the establishment of the principle of conservation of energy in all its generality. Thus the story of the emergence of the energy concept and the story of the establishment of the conservation law cannot be disentangled. There are several questions which the historian can ask in connection with the conservation law: as usual the question will be conditioned by one's presuppositions and bias. The most sophisticated question so far was the following: 'What are the reasons—scientific, social, political—that all the simultaneous discoverers of the principle made their discoveries in a short span of time in various places in Europe?' The span itself will be decided by what the questioner understands by 'discovery'. Generally the two decades 1840–60 will be meant. If one insists on Carnot in the group the time range becomes 1820–60. This question is sophisticated and modern in so far that it allows for both 'internalistic' and 'externalistic' answers; however it does not leave place for that unified approach which abolishes the internalistic-externalistic dichotomy, towards which historians of science are striving.

My approach will be somewhat different, as a result of which I shall be left with two questions instead of one. To neither of them do I have a simple and immediate answer. My thesis is that the principle of energy conservation is not a case of multiple discovery, but that in the span of 1840–60 different problems were bothering different groups of people in different places, and they came up with different answers. The answers turned out to be related, until finally in the 1860s they proved to be more than related, they turned out to be logically derivable one from the other.[3] By this I do not wish to imply that the problem is not to be solved in a social contect, but only that the model suggested for simultaneous discoveries is not the correct one.

Historiographical survey

To start on safe ground I shall attempt to formulate an unassailable first statement: in 1840 there was no principle of conservation of energy known. In 1860 conservation of energy was a victorious new achievement of science, with enormous impact not only on pure science, but on philosophy, literature and society in general. Hundreds of books and articles were written on the subject, mostly in popular magazines and public lectures. Out of this excitement grew the new science of energetics and several new branches of psychophysics. In the twentieth century to the best of my knowledge the first writer to have been interested in the 'discovery of the law of conservation of energy' was George Sarton. He describes shortly the work of Joule and Mayer and considers these two the original discoverers. Ther contributions are dealt with in a manner which could be called 'history-of-science-by-accumulation-of-flashes-of-genius' or simply 'Cleopatra's nose'—kind of history.

J. R. Mayer was a German physician with but little physical and mathematical knowledge. In the course of his service as a doctor aboard a Dutch ship he discovered the law of conservation of energy, by a sudden intuition. This great discovery, comparable in its suddenness to a religious conversion, occurred while he was in the harbour of Sarabaya (NE. Java) in July 1840.

As against Mayer 'who was primarily a philosopher' Joule, according to Sarton, was a 'metrologist':

His main interest lay in exact measurements and his special genius showed itself at its best in the invention of methods enabling one to obtain more and more accuracy in quantitative experiments.[4]

In the early thirties some books on history of science were published by the Marxist group in England.[5] Their work can be typified by the famous statement 'science owes more to the steam engine than the steam engine to science'. No specific treatment of the principle of conservation of energy was undertaken by them but in scattered records in books by Bernal and others it becomes clear that they considered the principle as a straight inductive result drawn from the accumulated experiences of steam-engines and other machines.

In the 1940s the social historian S. Lilley began publishing his important papers on the social aspect of the history of science, and especially his studies of heat. His approach is a Marxist version of the history of ideas. In his opinion, conservation of energy ideas were mostly *a priori*. The novelty in the 1840s was 'a new attitude of mind, a new habit of thought' which, however, was not caused by 'the internal growth of existing science but by the importation into science of habits of thought that were created by the social agency of the Industrial Revolution'.[6]

Elsewhere, dealing with Helmholtz's work I attempted to prove that the special coincidence of factors in mechanics, physiology, philosophical mood and availability of data which went into Helmholtz's work were present only in the Germany of the 1840s. Thus I certainly do not think as Lilley does, that:

Virtually all the scientific prerequisites for this discovery were in existence by 1800.[7]

Not only does this seem to be incorrect on my presuppositions, i.e. that the problem-situation had not yet been created in 1800, but even on Lilley's own theory, some important data were missing. Let me mention only one argument which applies to Helmholtz, Joule, Mayer, Kelvin and Rankine: they relied on the data supplied by the experiments of Regnault. Regnault did his measurements on the express order of the French ministry of Public Works and was financially supported by the ministry. His work involved high pressure studies, i.e. an experimental set-up which nobody in England or Germany could have afforded at the time.

An important work on the mechanical foundations of the conservation of energy principle is E. Hiebert's book[8] on this topic. As this treatise deals with mechanical energy only and follows the topic only up to the 1750s I shall not deal with it here. On the other hand, in a recent article, Professor Hiebert makes a few remarks on the principle, treating it as a case of multiple discovery:

Suffice it to say that conservation of energy was an independent, multiple discovery that burst forth among various European scientific investigators who were more closely tied to civil and military engineering, medicine, physiology, and brewing than to anything going on at the academic centers in the physical sciences.[9]

Two publications appeared recently on energy: a book by Theobald and an article by D'Haene.[10]

The most important work on the history of the conservation of energy principle is Thomas Kuhn's essay: 'Energy Conservation as an Example of Simultaneous Discovery'.[11] It appeared in 1959, and significantly, it was reprinted in a collection on the sociology of science, immediately following Merton's provocative 'Priorities in Scientific Discovery'.[12] Kuhn's main point is, that for various reasons which I shall analyse below, the time had come in 1842 that the principle of conservation would and should be discovered. As a result, Kuhn quotes twelve simultaneous discoverers of the law. The literature on simultaneous discoveries is rich and it has a tradition in sociology going back to the work of Ogburn and Thomas.[13] The underlying idea of this sociological treatment is summarized by Merton:

So the almost changeless pattern repeats itself. Two or more scientists quietly announce a discovery. . . . It is often the case that these are truly independent discoveries, with each scientist having separately exhibited originality of mind. . . . The situation is often ambiguous with the role of each not easy to demonstrate and since each knows that he had himself arrived at the discovery, and since the institutionalized stakes of reputation are high and the joy of discovery immense, this is often not a stable solution. . . . Then begins the familiar deterioration of standards governing conflictful interaction: the other side, grouping their forces, counter with the opinion that plagiary had indeed occurred, that let him whom the shoe fits wear it and furthermore, to make matters quite clear, the shoe is on the other foot. *Reinforced by group-loyalties and often by chauvinism, the controversy gains force, mutual recriminations of plagiary abound, and there develops an atmosphere of thorough-going hostility and mutual distrust.*[14]

Merton has added to Ogburn and Thomas's list of simultaneous discoveries, and according to the editors of *Sociology of Science*:[15]

Merton and Elinor G. Barber have, for example, compiled a list of more than 250 instances of independent multiple discoveries in science, each instance involving from two to fifteen or more scientists, often in several different countries.[16]

What is missing in all these works is a clear criterion of what shall be counted as a discovery. It is very probable that no such generalized criterion can be given, but what could have been pointed out is that at least in each specific case a criterion must be found. It is my feeling that had this been done, a good part of the 250 cases would turn out not to be cases of multiple discovery, but cases of unintentional collaboration. As I shall try to show for the case of the conservation of energy principle, the 'simultaneous discoverers' discovered very different things, and it is only under the influence of hindsight gained from their pooled results, that their discoveries seem identical. The various discoverers discovered different things because they asked different questions. Mayer discovered the indestructibility of forces of nature; Helmholtz discovered that the sum of the various kinds of force is a constant and that all must have the dimension of $m\mathbf{v}^2$; Joule discovered the mutual convertibility of heat and 'mechanical powers' as a result of which Thomson established the dynamical theory of heat on a firm foundation. Finally Clausius, Thomson and Rankine showed the equivalence of these various results and it becomes clear that one result is derivable from the other, i.e. that all of them had actually discovered the same thing. Coming now to Kuhn's paper let me say this: Kuhn's paper is not only the best treatment of our problem yet available, but also rich in unexploited suggestions and ideas which are only hinted at. This richness, however, is on the verge of becoming an embarrassment because no clear stand emerges. His very careful formulation almost contradicts his own title. He realizes that:

In the ideal case of simultaneous discovery two or more men would announce the same thing at the same time and in complete ignorance of each other's work, but nothing remotely like that happened during the development of energy conservation.

And more than that:

To each other at the time, they often communicated nothing at all. What we see in their works is not really the simultaneous discovery of energy conservation.

So why call it simultaneous discovery at all? Kuhn's answer to this implicitly formulated question is, that all of the pioneers contributed to the 'rapid and often disorderly emergence of the experimental and conceptual elements from which that theory was shortly to be compounded'. All this points to the fact that a criterion of 'discovery' has to be set up.[17] Now, as mentioned above, I do not expect Merton to have set up such a general criterion which would cover all of his 250 cases, because I do not believe that such a generalized criterion can be set up. But at least for each specific case some criterion must be found. The criterion I wish to suggest for this specific case (in the belief that it has a wider application than for this case) is that such discoveries should be considered simultaneous which give related answers to similar problems. As I shall show later there were at least two totally different problems which were solved respectively by Helmholtz and by Joule-Kelvin and Rankine. It needs much further research to decide what were the problems of the various minor figures in the story.[18] It should be pointed out however that when dealing with each individual Kuhn gives a penetrating discussion of the intellectual development of his subject, but he fails to generalize these analyses into a criterion of discovery by stressing the importance of the problem which any of the 'discoverers' set out to solve.

Different problems, different solutions

For us the principle of conservation of energy is an all-embracing principle of nature, with several other important conclusions considered as mere corollaries of it. Thus the mutual convertibility of the various kinds of energy is an equivalent statement to the principle of conservation. The mechanical nature of heat is a mere corollary from the conservation law. It is self-evident for us that phenomena of organic life are subsumed to the principle exactly as are the laws of inorganic nature. That this is so is a result of scientific proof of these equivalences, and to ascribe to it any historical truth is sheer hindsight.

Historically there were at least two independent and simultaneous developments in England and Germany between 1840 and, let us say, 1855. The English group, originating with Joule was preoccupied by problems of the efficiency of conversion between the various 'mechanical powers'. Their work resulted in the final proof that heat was a mode of motion. The Germans Mayer and Helmholtz were troubled by the physiological problem of 'animal heat' and their work resulted in the formulation of a law of conservation, in Helmholtz's case mathematically proved on the basis of correct dimensional analysis. Mayer believed in the caloric theory while both Mayer and Helmholtz emphasized that whether heat was matter or motion was not germane to the principle of conservation. The connecting link between the two trends is the work of Carnot to which all related and thus I shall review it briefly as a first thing.

The dynamical theory of heat

The old theory of heat was that of the mechanical motion. Bacon, Boyle, Newton, Locke all adhered to it more or less explicitly. On the other hand in the late eighteenth century and early nineteenth a consistent theory of chemistry which was fairly successful grew out of the work of Black and Lavoisier, relying on the caloric model of heat. The scientific value and inner consistency of the caloric theory has been demonstrated in the papers by Sanborn Brown and the earlier articles by S. Lilley.[19] I shall not go into this here. What is important, however, is that for his nineteenth-century readers it was beyond doubt that the caloric theory underlies Carnot's considerations in his *Reflections on the Motive Power of Fire*.[20]

There is little doubt in anybody's mind that Sadi Carnot was the man who discovered the Second Law of Thermodynamics, introduced the cyclical process which we call the Carnot cycle, and recognized the importance of reversible processes in general. So far no problems here, and his work is outside the scope of this study. However, there has been a long controversy going on in the last fifteen years between various physicist-historians on the historical and logical place of the First Law of Thermodynamics in the formulation of the Second Law. And this is intimately connected with this study. I intend to skip the 'internal history' of the second law, but will review briefly the controversy and take sides very definitely.

Sadi Carnot published his epoch making 'Réflexions sur la puissance motrice du feu' in Paris in 1824. In this paper (written, as noted by Kuhn, for a popular audience, for Carnot does not use any mathematics except in footnotes), besides the term 'chaleur' and 'feu' we find the term 'calorique'. With these concepts as his tools of thought he describes the problem facing him, namely the connection between the efficiency of a steam engine and the temperatures between which it works, and comes up with his theorem which is one of the numerous equivalent formulations of the second law.[21]

The model with which Carnot is working is clearly the caloric theory. He makes frequent comparisons between his engine working between two temperatures and the fall of a body from one height to the other, and clearly applies the law of conservation of matter to it. In view of this it seems to me indisputable that his 'calorique' is identical with Lavoisier's 'caloric'. Clapeyron, in his 'Memoir on the Motive Power of Heat' attributes to Carnot more than what Carnot ever said, but whatever Clapeyron attributes to Carnot is much less than what later historians of science thought that Clapeyron attributed to Carnot. All Clapeyron says is:

The idea which serves as a basis of his [Carnot's] researches seems to me to be both fertile and beyond question; his demonstrations are founded on *the absurdity of the possibility of creating motive or heat out of nothing.*[22]

If there is one central idea of Carnot's work it would seem to me to be rather 'wherever there exists a difference of temperature, motive power can be produced'. It is also in this connection, dealing with the two different temperatures that Carnot discloses in the clearest way that he thinks in 'caloric' terms:

The production of motive power is then due in steam engines not to an actual consumption of caloric, but to its *transportation from a warm body to a cold body.* . . . According to this principle, the production of heat alone is not sufficient to give birth to the impelling power; it is necessary that there should also be cold.

It is this statement which made Joule so convinced that Carnot had missed the whole point.

The very idea of 'equilibrium of caloric' with which all the above quoted passages are associated is, at this stage, not a concept from the motion theory of heat.

The historical complications began when in the 1911 edition of the *Encyclopaedia Britannica*, article 'Heat', H. Callendar suggested that Carnot's 'calorique' is actually the entity which later became known as 'entropy' (so called by Clausius), and that Carnot's model for heat was the motion-theory and not the caloric theory. This criticism was later taken up by several distinguished physicists like J. N. Bronsted, K. Schreiber, L. Brillouin and recently by Victor K. La Mer, and it has, I think, conclusively been answered by Thomas S. Kuhn.[23] After this basic statement, let me add a few qualifications. It is certainly true that Carnot mentions the impossibility of a perpetual motion machine, and takes this basic truth for granted. Again, as happened later in the case of Helmholtz, this in itself is not a sufficient condition for the formulation of a conservation principle—but it is a necessary one. Then, it is also true that Carnot often speaks in a way that can be interpreted as if he meant that heat was motion, thus a footnote we find:

The objection may be perhaps raised here, that perpetual motion, demonstrated to be impossible by mechanical action alone, may possibly not be so if the power either of heat or electricity be exerted; but is it possible to conceive the phenomena of heat and electricity as due to anything else than some kind of motion of the body, and as such should they not be subjected to the general laws of mechanics? Do we not know besides a posteriori, that all the attempts made to produce perpetual motion by any means whatever have been fruitless?

This passage is tell-tale (incidentally, La Mer quotes it too, without even mentioning that it is a footnote and not in the body of the text)l It is a methodological comment, not a scientific one. Carnot surely read the Lavoisier-Laplace joint enterprise, and it is very probable that he was influenced by the methodological aspect of it. It is an ever-recurring pattern that scientists pay lip-service to a methodology which partially or even completely belies their own conceptual framework. In this case what I want to argue is that Carnot was not an ardent, fanatical believer in the material theory of heat even at this time, and when pressed by a methodological problem he was very much aware of that; but at this stage of his work he thought in caloric terms. When dealing with science proper he betrays this; I shall quote another short passage which sounds even more 'caloric-minded' than the one quoted so far: the reason I did not bring this one in when talking about strong evidence, is that it is so typically unintentional:

The caloric developed in the furnace by the effect of the combustion traverses the walls of the boiler, produces steam, and in some way incorporates itself with it.

If we try to unite these two approaches we will have to say simply that Carnot was in reality undecided. In the original Memoir he says:

For the rest we may say in passing, the main principles on which the theory of heat rests require the most careful examination. Many experimental facts appear almost inexplicable in the present state of this theory.

The philosophical status of this undecidedness is very different from a non-commital attitude on methodological grounds. It is not that Carnot does not believe that there is a 'real' solution to the problem, and that he considers theories operationally—it is only that he does not know; and this is a great difference indeed! That this reinterpretation is false has been shown by Kœnig and Kuhn, furthermore: 'entropy' cannot systematically be substituted for 'calorique': very often we would get complete nonsense.[24]

Carnot died very young; his brother H. Carnot submitted a 'Lettre' to the Académie des Sciences. In this letter he states that the accompanying unedited fragments of his brother 's'ils n'apportent point à la Science des résultats nouveaux, témoignent que Sadi Carnot avait prévu avec une assez grande netteté ce que l'on a plus tard tiré de ses idées'. Very probably here originates the later myth that Carnot had from the beginning been in full possession of the foundations of modern thermodynamics. (At least La Mer relies on the posthumous papers too.) Whether this was implied by his brother or not, I do not know, but it is certainly false. However, these notes do contain serious changes in Carnot's conceptions; there is a clearly formulated switch to the motion theory of heat, and an almost clear formulation of conservation of energy principle. When the notes were written, is not clear; in Mendoza's selection they are arranged according to the dating of Raveau; but as Carnot died in 1831, at the age of thirty-six, it is not very important to us to try to settle this question here. Had this been a study in priorities, this point would have necessitated a profound analysis. But here it is enough to remind ourselves that these notes lay unnoticed till 1878. It is certainly true, that having arrived at these conclusions later, it is more understandable that La Mer and others tried to read these thoughts back to the original Memoire. I certainly agree with Kuhn that 'the notes reflect on the brilliance of the memoir's author, not on the memoir itself. Since they refer to problems developed in the memoir, and are occasionally in fundamental conflict with it, most, or all of the notes must have been written after 1824'.

From these notes it becomes clear that Carnot knew very well the work of Rumford and Davy, and moreover, in this, he is nearer to the position of Helmholtz twenty years later, he knew the work that had been done in mechanics. He knew the conservation law of *vis viva* and he certainly understood the generality of the mathematical treatment involved. Had he been absolutely sure, even now, in the motion-theory of heat, and had he lived to perform the experiments he suggested for testing the heat-work equivalence, he might have founded thermodynamics.

Joule, Thomson, Rankine

Joule was led to his measurements of the mechanical equivalent of heat through his interest in improving the electro-magnetic machine—what we call the dynamo. He was interested in increasing the lifting force of the engine and his question was: how will a given battery obtain the greatest amount of force out of the least weight of soft iron and wire. This at once introduced him to the measurement of the physical actions in his engine in terms of 'work', the only measure of mechanical effect.

In his second paper on this topic in the form of a letter to the editor of *Sturgeon's Annals of Electricity* he says about his new engine:

I finished the one I was working at during last summer; it weighs $7\frac{1}{2}$ lb and the greatest power I have been able to develop with a battery of forty-eight Wollaston four-inch plates was to raise 15 lb a foot high per minute, in which estimate the friction of the working parts, which was very considerable, was reckoned as the load.

Here he uses 'power' in the usual sense, and the passage shows Joule's realization of the importance of this fundamental concept.

Yet even at this stage he is still far from realizing the connection between the work which his machine can perform and the source of the work, namely the consumption of metal, in this case zinc, in his battery. This early paper ends indeed with an announcement which shows almost an acceptance of perpetual motion.

I can hardly doubt that electromagnetism will ultimately be substituted for steam to propel machinery. If the power of the i's in proportion to the attractive force of its magnets, and if this attraction is as the square of the electric force, the economy will be in the direct ratio of the quantity of electricity and the cost of working the engine may be reduced ad infinitum.

This sanguine hope, however, brought disappointment. The series of papers ends with the realization that the growing speed of the machine induces resistance in the magnet and finally that no electro-magnetic machine could compete with the Cornish engine. But what he learned from the research was to measure work and electricity in absolute units.

The next steps followed in rapid succession. He discovered the connection between the current and the heat produced; the heat evolved during the electrolysis of water; his previous research taught him the mechanical power of current, and now he realized: 'the mechanical and heating powers of a current are proportional to each other'. At this stage it becomes crucial to decide if heat is destroyed and something else generated, or is it true, as the caloric theory explains, that the matter of caloric merely transmits something on its way between two parts of the machine which 'something' can turn into mechanical power. If so, what is this something? In his address to the meeting of the British Association at Cork in 1843 Joule says:

I resolved therefore to endeavour to clear up the uncertainty with respect to magneto-electrical heat.[25]

In order to test his preconceived idea,[26] that indeed heat is motion he does an unusual experiment: he encloses the armature of an electromagnet in a container filled with water. The armature revolves in this. The container is cylindrical in shape and it is carefully insulated to prevent heat losses. He takes the temperature and then rotates the whole container between the poles of a stationary electromagnet. The induced current is measured and also its temperature is again taken. The quantity of heat evolved directly in the revolving coil is calculated. This was a proof that heat was not just something transferred by the caloric in the current from one place to another but that it was actually *generated*! Joule has proved to his complete satisfaction the dynamical theory of heat, and now he could face the next problem: if so, then there must be a 'mechanical value' of heat which must be established. This he went on to do, and we all know how brilliant his work was. Thus Joule by 1843 had established the dynamical theory of heat, and found the first value of the mechanical equivalent of heat. That is what Joule discovered. What he did not see was that he had discovered the conservation of energy, i.e. the fact that the sum of all kinds of energy in a system is a constant. What we do find in his papers are a few very near misses. In a postscript to the Cork address he says:

I shall lose no time in repeating and extending these experiments, being satisfied that the grand agents of nature are, by the Creator's fiat, *indestructible*; and that whenever mechanical force is expended, an exact equivalent of heat is always obtained.[27]

It seems that his conserved entity is nearer to a rather 'vague' force than to our 'energy', very much in the Faraday tradition.

Joule's exposition closest to the principle was included in a popular lecture at St Ann's Church Reading Room in Manchester on 18 April 1847. The lecture was called 'On Matter, Living Force, and Heat'[28] and its central idea is that all forms of energy are convertible to heat. We should keep in mind that the concept of energy, let alone kinetic and potential energy in our modern sense were not then part of the scientist's vocabulary. Here he says:

The common experience of everyone teaches him that living force is not *destroyed* by the friction or collision of bodies. We have reason to believe that the manifestations of living force on our globe are, at the present time, as extensive as those which have existed at any time since its creation, or, at any rate, since the deluge—that the winds blow as strongly, and the torrents flow with equal impetuosity now, as at the remote period of 4,000 or even 6,000 years ago; and yet we are certain that, through that vast interval of time, the motions of the air and of the water have been incessantly obstructed and hindered by friction. We may conclude, then, with certainty, that these motions of air and water, constituting living force, are not *annihilated* by friction. We lose sight of them, indeed, for a time; but we find them again reproduced. Were it not so, it is perfectly obvious that long ere this all nature would have come to a dead standstill. What, then, may we inquire, is the cause of this apparent anomaly? How comes it to pass that, though in almost all natural phenomena we witness the arrest of motion and the apparent destruction of living force, we find that no waste or loss of living force has actually occurred? Experiment has enabled us to answer these questions in a satisfactory manner; for it has shown that, wherever living force is *apparently* destroyed, an equivalent is produced which in process of time may be reconverted into living force. This equivalent is *heat*. Experiment has shown that whenever living force is apparently destroyed or absorbed, heat is produced.[29]

Allow me to risk tedium, and to emphasize again: Joule talks about the general convertibility of all forms of 'forces' into heat. He does not say anywhere that these various forces are actually the same kind of thing, or that their sum is constant. Nor does he apply dimensional consideration. That what Joule said was equivalent with let us say Helmoltz's formulation was proved only in 1855 by Rankine.

Joule did not know about the work of Carnot. It was only in the Oxford meeting of the British Association, when Joule repeated his results that he met William Thomson who knew about Carnot's work from Clapeyron's paper and pointed to the contradiction between the two theories. Joule's and Kelvin's versions of this memorable meeting were reprinted by Reynolds and by Watson. What we know about the Oxford meeting is that Joule was told by the chairman to cut his paper short and to stick to experimental details. We know that his paper made little impression, and that Kelvin quotes Graham's and Miller's objection to Joule's results on the ground that the results depended on fractions of a degree of temperature—sometimes very small fractions. The historian's sceptical mind is not satisfied with the usual explanation that Joule's degree of accuracy was unusual in the science of his days; rather Graham's and Miller's rejection (actually these two are only typical of the profession at the time) prove that mental attitudes were not yet ready for the new ideas.[30]

But not only Miller and Graham rejected Joule's results. Thomson, the first to have recognized the importance and significance of Joule's results, though greatly impressed, was not convinced. The stumbling block was Carnot's theory.

Kelvin's primary interest in the Carnot-Clapeyron theory was his striving for the establishment of an absolute scale of temperature—namely, a thermometer which is independent of thermometric substance. This is feasible on the Carnot theory for here 'quantity of heat and intervals of temperature are involved as the sole elements in the expansion for the amount of mechanical effect to be obtained through the agency of heat'.[31]

It is precisely here that Thomson's scientific insight is seen. He set himself a new problem: to remove the contradiction between the results of Joule and Carnot. That Carnot had valuable and experimentally proved results he knew. That Joule had highly accurate results he saw too. How was it possible that their correct results were derived f:om contradictory hypotheses? The next few years were spent on this problem. Finally, in 1851 Thomson had the answer. His epoch-making paper appeared in the Transactions of the Royal Society of Edinburgh. Its title is 'On the Dynamical Theory of Heat'.

Considering the problems of Joule and Thomson, it is clear that it was absolutely essential to decide between the caloric theory of heat and the dynamical theory of heat, that is between a conservation principle and a convertibility principle. Thus it is no wonder that this 1851 paper, where he successfully reconciles the Joule and Carnot principles, is not called 'on the conservation of . . .' but is called '*on the dynamical theory of heat*'.[32] It is, however, important that when writing his article Kelvin did not yet know of Helmholtz's paper, and thus his article does not mention the conservation of energy as such. Far from it. At the time Kelvin's thoughts were occupied by a different topic. If indeed heat is not lost on either theory—Carnot's or Joule's—there is something that is lost. It was as a result of his attempt to reconcile the two principles that he arrived at the solution of this problem: the entity which is lost is not energy, but concentration of energy—this is the great generalization expressed as the 'universal tendency in nature to the dissipation of energy'. This was so important for him that he even suggested a new name for it: *motivity*.[33]

Rankine is the third major figure in the English group. He too has different problems to solve but his problems also demand a clear cut decision whether to adopt the caloric theory or the mechanical theory of heat. Rankine was primarily interested in molecular constitution of matter. He was trying to develop mathematical models for the elasticity of gases and vapours and this brought him to a consideration of the mechanical action of heat. He accep:ed Joule's view of heat at an early stage mainly because he assumed a 'Hypothesis of Molecular Vortices', according to which:

quantity of heat is the vis viva of the molecular revolutions or oscillations.[34]

In this work Rankine develops mathematically the foundations of the dynamical theory of heat and quotes Joule's 1845 paper 'On the Changes of Temperature Produced by the Rarefaction and Condensation of Air'. Here Joule introduced a kind of vortex hypothesis in which Rankine in all probability followed him. Rankine however does not give Joule any credit on two important points where Rankine's research concentrated: determining the absolute zero of temperature and the law of specific heats. Both these were mentioned by Joule in articles published before the above-mentioned 'The Changes of Temperature . . .' and were not mentioned in this one. We can only assume that Rankine did not know about the earlier papers.

From our point of view the important thing about the early work of Rankine is that he, relying on a hypothetical model of the constitution of matter arrived at general equations relating pressure, volume, temperature and heat. These equations implicitly conform to the law of conservation of energy and to Carnot's principle, yet Rankine does not draw a general conclusion either about the conservation of energy nor drawing attention to the importance of the mechanical equivalent of heat. His achievement was mathematical, with no clear interpretation of his own results.

Though it is not worth going into detail here as to his equations, it is important to understand his model of the constitution of matter. The vortex hypothesis is explained in two papers both published in 1850.[35] Three presuppositions underlie this theory of matter; the first is:

Each atom of matter consists of a nucleus or central physical point, enveloped by an elastic atmosphere, which is retained in its position by forces attractive towards the nucleus or centre.

His second presupposition is:

That the elasticity due to heat arises from the centrifugal force of revolutions or oscillations among the particles of the atomic atmospheres; so that the quantity of heat is the *vis viva* of the revolutions or oscillations.

Finally, Rankine also introduces a third hypothesis which is supposed to connect the second hypothesis with the undulatory theory of radiation:

That the medium which transmits light and radiates heat consists of the nuclei of the atoms, vibrating independently, or almost independently, of their atmospheres.[36]

The first hypothesis is considered by Rankine as the fundamental assumption of the theory of matter, of which, as we have seen above, the old motion theory of heat is a simple corollary. One will not be able to detect in this important work even an adumbration of the principle of conservation of energy. It is an interesting rare case where the mathematical sophistication (an advanced mathematical theory of heat is developed here) is far ahead of the philosophical interpretation of the results.

Thus the result of the Joule-Thomson-Rankine collaboration was a well-established dynamical theory of heat. Having started from different problems, all or which, however, necessitated the clarification of the nature of heat, they came up with slightly different but related results. Joule measured with the greatest accuracy the mechanical equivalence of heat. He knew that wherever mechanical power seems to disappear, it will reoccur in the form of heat. Thomson, who was the first to realize that Carnot's theory is in contradiction with Joule's results decided to clear up the confusion. At first he declared that 'the conversion of heat (or caloric) into mechanical effect is probably impossible, certainly undiscovered'. Later he took Joule's results seriously, and asked whether Carnot's results could be saved without Carnot's assumption. His greatest achievement was that he found how to modify Carnot's results so that both Joule's theory and Carnot's insights could be saved. His great paper 'On the Dynamical Theory of Heat' begins with the two fundamental laws together:

the whole theory of the motive power of heat is founded on the two following propositions, due respectively to Joule and to Carnot and Clausius.
    Prop. I (Joule)—When equal quantities of mechanical effect are produced by any

means whatever from purely thermal sources, or lost in purely thermal effects, equal quantities of heat are put out of existence or are generated.

Prop. II (Carnot and Clausius)—If an engine be such that when it is worked backwards the physical and mechanical agencies in every part of its motions are all reversed, it produces as much mechanical effect as can be produced by any thermodynamic engine with the same temperatures of sources and refrigerator, from a given quantity of heat.[37]

These two principles were later to be called the fundamental axioms of thermodynamics—the new name given by Rankine to the dynamical theory of heat. But in this formulation the first principle is not yet identical with the principle of conservation of energy. Thomson had not yet read Helmholtz's 'Über die Erhaltung der Kraft',[38] and he nowhere mentions conservation of energy.[39] As we have seen, until 1852 Rankine himself had not yet come up with the principle of the conservation of energy. The English groups had established the dynamical theory of heat and the principle of convertibility.[40]

The conservation of energy

The second trend of development went a different way. In this group I consider all those 'discoverers' who were preoccupied with conservation ideas, and whose problems were formulated in terms of conservation or loss. The background to their problem was sometimes purely metaphysical—as in the cases of Faraday and Ørsted, sometimes physiological,[41] and sometimes a combination of both, as in the cases of Mayer and Helmholtz. The most successful in this group was Helmholtz, and he was the man who gave the standard formulation in mathematical terms to what has become known as the principle of conservation of energy.

As I have dealt at great length on the work of Helmholtz elsewhere,[42] I shall only summarize the main points about his work.

Helmholtz combined in his personality, education, and the fact that he grew up in Germany, all the factors which were a solid and necessary basis for the enunciation of the conservation principle. These were:

(1) An *a priori* belief in general conservation principles in Nature.
(2) Realization that it is not enough that two formulations of[43] mechanics: the vectorial-Newtonian and scalar-analytical-Lagrangian, are mathematically equivalent, they must also be conceptually correlated.
(3) An awareness of the physiological problem on 'animal heat' or more generally of 'vital forces', and a belief that these are reducible to the laws of inanimate nature.
(4) A mathematician's certainty[44] that whatever is the entity which is conserved in Nature it must be expressible in mathematical terms, and a mathematician's skill to perform the task.

He had read at a very early age the works of Newton, Euler, d'Alembert and Lagrange (though not Hamilton), and was aware of the double tradition in mechanics, and that something had to be done about it. It was clear to him that the central concept in Newtonian-vectorial mechanics was the concept of force, and at the same time he saw that the only quantity conserved in scalar-Lagrangian mechanics was the sum of '*vis viva*' and the 'potential function'. By temper and intellectual heritage he was a disciple of Kant and thus committed to a belief in the great unifying laws of nature; this took the form of conservation laws, and naturally the conserved entity had to be that vaguely defined entity 'Kraft' or 'force' (in the Faraday sense).[45] All this was in complete harmony with his mechanical philosophy: a belief that all phenomena of Nature are reducible to the laws of mechanics. By training he was a physician and he spent several years in the laboratory of the famous physiologist Johannes Müller. There he came to face the problem of 'vital forces' and especially that of animal heat, and his first works were in this field. Again his approach was that 'vital forces' are like other forces, conserved in Nature and as all phenomena are reducible to mechanics, so 'vital forces' must be reducible to mechanical forces. On top of all that Helmholtz was a mathematician of the first rank. He saw very clearly that if 'Kraft' is conserved in Nature, and mechanical energy is conserved in mechanics, then all 'Kraft' must have the same physical dimension as mechanical energy and must be, moreover, reducible to it. That is exactly what he did in his 1847 paper. The combination of philosophical, mathematical and physiological background was very typical of Germany of the 1840s. In England most of the scientific activity concentrated in the scientific societies which had a strong bend towards the practical side of science, and around the new industrial centres. Indeed, one of the shapers of nineteenth-century England, Benjamin Disraeli, wrote in *Coningsby* in 1844: 'It is the philosopher alone who can conceive the grandeur of Manchester and the immensity of the future.'

Paris, in the first half of the nineteenth century, was the scientific capital of the world, and the seat of the government was in the Ecole Polytechnique. Here the greatest mathematicans of the age studied, taught and created. This was the country where chemistry had been reformulated by Lavoisier a decade before, where Newtonian science had been brought to the zenith by Lagrange and Laplace, where the new mathematical foundations of a theory of heat were laid down by Fourier, Cauchy and S. Carnot. French science was concentrated in one place, in Paris, exercised by individuals who were creative in one special field, founded no 'school' and perhaps with the sole exception of Auguste Comte, did not have a philosophical system.

The situation was very different in Germany. This was the country where the great philosophical systems of Kant, Fichte, Schlegel, Hegel and later Lotze were conceived, at a time where in the other two countries nobody tried to construct a complete philosophical system. But even more important than that, a vast network of universities was founded, with a rigid hierarchy between them, where philosophical faculties were established, and where the philosophers expounded their systems. The broad background in these universities was such that whatever the student studied he could not have avoided facing sooner or later the great metaphysical problems posed by the various 'Weltanschauungen'; while in England 'science' was pursued, or in the traditional spirit 'natural philosophy' was taught, German 'Philosophie' covered the whole of the human intellectual enterprise. Speculation was encouraged; even today, after the great, late-nineteenth-century battle to erase the last remnants of an influence of the 'Naturphilosophie', or rather of a degenerated, ridiculously trimmed-down version of it, 'speculation' does not cause such a contempt in German as it does in English. It was not only the rigid, much ridiculed German 'Geheimrat' system, which made people speak of the 'school' of one or the other of the great philosophers or scientists. It was rather the fact that indeed these 'schools' or 'laboratories' represented a complete philosophical system, and every student had to take a stand towards them. One could not have worked in Weber's physical laboratory, or in Liebig's laboratory, without taking a deeply considered philosophical approach to Kant's epistemology or to the question of the mechanistic-vitalistic controversy. In this atmosphere one could not very well separate experimental data from highly speculative hypotheses.

Far be it from me to claim that this in itself is good or bad: what I do claim is that exactly this atmosphere which was detrimental for many other scientific problems (as for example for the first formulations of electrodynamics: indeed all German theories failed while the English, French and the isolated Ørsted did work) was indispensable for the establishment of the conservation of energy principle. In these universities the 'schools' of Liebig, Woehler, Johannes Müller, Weber and Gustav Magnus were founded. And here it was that the further prerequisite—namely, the cross fertilization between physical and physiological thought—was made possible.

Helmholtz's father was a close friend of Fichte's son, himself a professor of philosophy, and Helmholtz himself, according to his own testimony, was deeply influenced by both Kant and Fichte. In addition to this, Helmholtz's education in physics and his mathematical ability made him the ideal man for the task awaiting him.

The scientific problem on which Helmholtz worked was the origin of 'animal heat', i.e. the source of the vital forces. He believed in conservation of forces and was committed strongly to the reductionistic belief that all vital phenomena must be reducible to the laws of physics and chemistry. He faces the problem in his early physiological papers[46] and gives the answer in his famous 1847 'On the Conservation of Force'.[47]

The important thing for the point of view taken up here is that while Helmholtz proves clearly that the sum of tension forces (our potential energy) *vis viva* and heat must be a constant, he nowhere definitely relies on the dynamical theory of heat or does he assume it. True, he weighs the two theories and prefers the dynamical theory but this is neither his problem nor an integral part of his answer.

Helmholtz poses two questions:

In those cases where the molecular changes and the development of electricity are to a great extent considered, the question would be, whether for a certain loss of mechanical force a definite quantity of heat is always developed, and how far can a quantity of heat correspond to a mechanical force.[48]

There are two different questions for him. He quotes Joules' letter to the editor of *Philosophical Magazine*[49] where in Helmholtz's opinion answer is given to the first question. For the second question—namely, 'how far can a quantity of heat correspond to a mechanical force' he weighs the caloric against the dynamical theory but does not decide unambiguously between them.

To sum up, Helmholtz showed mathematically the truth of the principle of the conservation of energy, and that was the solution to his problem.

In order to see the synthesis of the two different results we now have to return to Rankine.

On 5 January 1853 Rankine read a paper before the Philosophical Society of Glasgow, called 'On the General Law of the Transformation of Energy'.[50] Here a correct and modern formulation of the principle of the conservation of energy appears without however quoting any authority or previous work.[51] Rankine was generally very scrupulous in his acknowlegements and very much aware of any kind of innovation. How then is one to esplain this omission? One possible explanation is that in Rankine's mind as in the minds of Joule, Thomson, Helmholtz and Clausius this principle suddenly appeared as self-evident when the dynamical theory of heat was established in a final form, and the seeming discrepancies between Joule's and Carnot's theories were removed.

However, we can trace how his conviction grew. A year before the above-mentioned paper, Rankine read another one before the British Association at Belfast on 2 September 1852. Here he speculates on a cosmological hypothesis which would avoid the disaster of the 'Wärmetod' (heat-death): one of the consequences of the second law of thermodynamics which predicts a slow dissipation of the energy of the world which will result in an equalization of temperature in the universe—so called death. Here he says:

The experimental evidence is every day accumulating of a law which has long been conjectured to exist—that all the different kinds of physical energy in the universe are mutually convertible; that the total amount of physical energy, whether in the form of visible motion and mechanical power, or of heat, light, magnetism, electricity, or chemical energy, or in other forms not yet understood is unchangeable. . . .[52]

Significantly Rankine mentions the convertibility and the conservations as two different experimentally verified principles. The essential unity of these laws must have been suddenly triggered in his mind some time between these two lectures. It is two years later, in 1855, when Rankine published his 'Outlines of the Science of Energetics'[53] that the situation becomes clear. He talks there of two axioms. One is a result of Joule's and Rankine formulates it in three equivalent forms:

All kinds of energy are homogeneous.

Any kind of energy may be made the means of performing any kind of work.

Energy is transformable and transferable.

The second axiom is:

The total energy of a substance cannot be altered by the mutual actions of its parts.

Then Rankine adds that it follows from the second axiom that 'all work consists in the transfer and transformation of energy alone'. It seems to me that if I read Rankine here correctly he is actually pointing out that the two axioms are identical. The two lines of development have united. The general principle of the conservation of energy with the dynamical theory of heat and the mutual convertibility of all kinds of energy as its corollaries has been established.

It took twenty more years until the general public realized this. The first popular book in England to be called *Conservation of Energy* was Balfour Stewart's in 1874. This leaves me with the two highly significant historical questions unanswered; allow me to repeat them (in spite of note 3):

(1) Why was it that all the diverse and seemingly unconnected elements that went into the conservation principle, were discovered in the mid-nineteenth century?

and

(2) What are the factors which created a social mood which was not only receptive to all these new ideas, but helped to unite them into a new principle of nature and developed into almost a new religion—the religion of energetics?

I gratefully acknowledge my general indebtedness to my friend Dr Arnold Thackray for many interesting and fierce discussions. This indebtedness applies also to a seminar on the 'Social Background of Science' run jointly by Professor E. Mendelsohn and Dr A. Thackray at Harvard University in 1968.

NOTES

1. It would be natural to include 'motion' among these central concepts. I left it out, because between the days of Descartes and the late nineteenth century, e.g. Maxwell, the concept of motion, or that of mobility, oscillated between being considered a primary and a secondary property of matter. On the other hand it was 'force' that replaced 'motion' for Leibniz and, though less consciously, also for Newton.

2. Planck, *Das Prinzip der Erhaltung der Energie*. This book was written in 1887 as a competition essay for the faculty of philosophy at Göttingen. The purpose of the competition was to find out whether that which William Thompson called 'conservation of energy' is identical with Helmholtz's 'conservation of force'. Planck's book is one of the most extensive treatments of the history of the conservation principle and also a detailed exposition of its importance for science and of its philosophical meaning. It is interesting to note that Planck in 1887 still finds it necessary to indulge in the inductivist-postivist witch-hunt against the 'Naturphilosophie'; on the other hand his insights and explanations are far beyond any methodological school. The quotation I am referring to above is:

I shall deal with the concept of energy only so far as it can be connected with the principle, presupposing that the concept of energy gains its meaning in physics first of all through the principle of conservation, which contains it [page v, my translation].

Poincaré is much more explicit on this subject, he ends his discussion by saying: There only remains for us one enunciation of the principle of the conservation of energy: there is something which remains constant.

It is this something which is called energy. Naturally after that there is no more place for experimental proof at all:

> Under this forum it is in its turn out of the reach of experiment and reduces to a sort of technology [in *Science and Hypothesis*, ch. VIII].

3. I am fully aware that this leaves two questions to be answered:

(1) Why was it that all the diverse, and seemingly unconnected elements that were later united under the blanket-law of conservation of energy, were discovered in the mid-nineteenth century? and

(2) What are the factors which created a social mood which was not only receptive to all these new ideas, but helped to unite them into a new principle of nature and developed into almost a new religion—the religion of energetics?

As my paper is devoted to a brief historiographical survey of the historical work on the conservation principle, and to a partial proof of my thesis, I shall not deal here with these questions, so let me suggest here a tentative answer to them.

The tentative answer to question one is; in the first half of the eighteenth century several trends ran parallel:

(*a*) Au tilitarian interest in machines brought to a search of how to make their performance better. This led to research which had to come up with a final answer whether the caloric theory or the dynamical theory is the 'true' one.

(*b*) A philosophical attitude brought attention to various forces of nature, their mutual convertibility and conservation in a vague sense.

(*c*) Physiological research had reached a point where the question had to be faced whether vital phenomena are subsumed to the laws of physical nature or not.

(*d*) Developments in mathematics, especially in rational mechanics brought the workers in the field to realize that mathematical equivalence must be proved between the vector mechanics developed by Newton and his English followers, and the scalar mechanics developed on the continent by Euler and Lagrange. This also resulted in the growing awareness of the importance of dimensional analysis: forces which are philosophically or physically convertible to each other must have the same physical dimensions.

The various discoverers having different problems, embodied in themselves different combinations of these trends and accordingly came up with different answers. About this later.

The tentative answer to the second question is:

The socio-political developments in Europe in the nineteenth century were such that by the 1860s a new religious attitude was ripe and looking for a theme. Much of institutional religion had been discredited by science and after Darwin this became even stronger, and the ground was prepared for a new belief in great all-uniting principles of nature: this was supplied by the principle of the conservation of energy. There is something stable and unchanging in the universe, combining organic with inorganic, terrestrial with the cosmos. I realize full well how tentative these answers are, and how much historical research has to go into substantiating them.

4. George Sarton. 'The Discovery of the Law of Conservation of Energy'.

Comparing the various stages in the development of the treatment of any problem in the history of science shows the enormous distance between the historical naïveté of the above quotations to the sophistication of, let us say, Kuhn's treatment of the problem as one of 'simultaneous discovery'.

5. I have in mind works like Bernal's *Social Function of Science*; Hessen's *On the social and economic roots of Newton's Principia*; B. Farrington's work on the Greeks and on Bacon and the various articles by Edgar Silsel.

In a recent study of Helmholtz in Russian by Lebedinskii and colleagues (1966) the conservation of energy is mentioned only in a few lines. The idea is that no genius was necessary for that, it being a simple inductive inference.

6. Sam Lilley. 'Attitudes to the nature of heat about the beginning of the century', *Arch. Intern. Hist. Sci., 28* (1948), p. 630.

7. Lilley, 'Cause and Effect in the History of Science,' from where the above quotation is taken.

8. Hiebert, *Historical Roots of the Principle of Conservation of Energy.* The relevance of this book to the developments in the nineteenth century is demonstrated in Hiebert's critical commentary to Kuhn's paper (*Critical Problems in the History of Science,* ed. M. Clagget, pp. 391–8).

9. Hiebert, 'The Uses and Abuses of Thermodynamics in Religion,' *Daedalus Fall,* 1966. p. 1046. Whatever will be said below about Kuhn's treatment of the problem as one of simultaneous discovery will apply here too. I shall return later to the other serious claim of Professor Hiebert which is the second part of the quotation.

10. Both appeared after Kuhn's paper, but because of Kuhn's special point of view, I shall mention these two first. D'Haene's point of view is mainly philosophical. He sees a fundamental difference between conservation of force and conservation of energy. The first is a metaphysical truth while the other is an experimental result. Theobald's book is a careful discussion of the status of the concept of energy in various branches of physics. Though rich in historical documentation the book is not primarily a historical study.

11. In Clagett (ed.) *Critical Problems in the History of Science,* pp. 321–56, and also in Barber and Hirsch *Sociology of Science,* pp. 486–515.

12. In Barber and Hirsch (eds.), *Sociology of Science,* pp. 447–850. There is another important essay by Merton on essentially the same problem: 'Singletons and Multiples'; also, 'Resistance to the Systematic Study of Multiple Discoveries in Science'.

13. In the article 'Are Inventions Inevitable?' Here the authors collected some 150 cases of what they considered cases of independent discovery, or, as Merton calls them, 'multiples'.

14. Merton, in Barber and Hirsch, *Sociology of Science,* p. 175, note 15.

15. Ibid., p. 444, note 14.

16. Let me mention a parallel case here. There is an old discussion in the history of science on the priority (a 'Prioritätsstreit') of Boyle, Townley, Powers and Hooke on 'who discovered Boyle's law?' The various authors on the subject presuppose that there is one law, i.e. 'Boyle's law' which was to be discovered. In a recent article J. Agassi points out that there were two different laws: Boyle discovered the gas law for compression, that is for pressures above one atmosphere, while Townley discovered a law for rarefaction, that is for pressures below one atmosphere. It was this law which Boyle openly acknowledged to Townley and on which the whole priority debate is based.

17. How badly a criterion is needed can be illustrated by the following example: Kuhn recognizes Holtzmann's claim as one of the simultaneous discoverers because to several of his contemporaries 'Holtzmann seemed an active participant in the evolution of the conservation theory', although in Kuhn's own opinion: 'Holtzmann can scarcely be said to have caught any part of energy conservation as we define that theory today.' So it seems that it is the contemporary judgment which serves as a criterion. Yet immediately after Holtzmann, Kuhn deals with Hirn and recognizes his claim on the merit of his work, although 'none of the standard histories cites these articles or even recognizes the existence of Hirn's claim'. Somewhat later Kuhn adds that had Hirn's work been known he surely would have been regarded as one of the contributors.

18. The importance of problems as against the solutions for the history of science has often been pointed out by Popper and his followers. For an explicit study of this see J. Agassi: 'The Nature of Scientific Problems'. J. Ravetz is working on a study of problems in history of science. In a recent article 'The Evolution of Science and its History', he says:

> Seeing science as the investigation of problems, rather than the discovery of facts or truths enables us to see science as a creative endeavour conditioned by its environment in a host of direct and subtle ways.

19. 'The Caloric Theory of Heat', *Amer. J. Phys.*, *18* (1950), pp. 307–73. S. Lilley, 'Attitudes to the Nature of Heat'.

20. In Mendoza (ed.), *Reflections on the Motive Power of Fire*.

21. In H. M. Evans (ed.), *Men and Moments in the History of Science* (Seattle, 1959), pp. 57–111.

22. Mendoza, *Reflections on the Motive Power of Fire*, p. 74.

23. The papers of La Mer and Kuhn in their order of appearance are: La Mer 'Some Current Misrepresentations', Kuhn, 'Carnot's Version of "Carnot's Cycle" ', and La Mer 'Some Current Misinterpretations II'. The second paper by La Mer is intended as a refutation of Kuhn's paper, but it seems to me that it does not succeed in invalidating Kuhn's argument. The much earlier essay by Koenig is a very profound analysis of Carnot's work, and in many respects anticipates and answers La Mer's criticism.

24. Elsewhere I have given a similar argument with respect to Helmoltz's 'Kraft', and there I go into detail to prove the falsity of the 'substitution' approach.

25. J. P. Joule, *Scientific Papers*, I, p. 124.

26. In an 1844 Appendix to one of his earliest papers, Joule said:

> I have once or twice made use of the terms 'latent heat', 'caloric', etc.; but I wish it to be understood that these words were only employed because they conveniently expressed the facts brought forward. I was then [i.e. January 1843] as strongly attached to the theory which regards heat as motion among the particles of matter as I am now. [*Scientific Papers*, I, p. 121].

27. ibid., p. 158.

28. This is the source which is generally considered when Joule's discovery of the principle is advocated. (See, for example, E. C. Watson, 'Joule's Only *General* Exposition of the Principle of Conservation of Energy', *Amer. J. Phys.*, *15* (1947), p. 383.) What Watson and others do not ask is, how is it possible that if Joule realized this implication of his discovery he never wrote on that topic again, nor did he claim the 'conservation of energy principle' as his own contribution, nor do we see any internal change in his work after this alleged 'General exposition'. The idea that Joule is the discoverer of the principle goes back to the classical biography by Osborn Reynolds: 'Memoir of J. P. Joule'. (4), *6* (1892), pp.1–191.

On the other hand the only other book on Joule of which I know, Alex Wood's, *Joule and the Study of Energy*, does not claim for Joule the establishment of the conservation principle, but rather, in the same line which I shall take up here, he says:

> The honour of having established the mechanical nature of heat upon a firm foundation belongs unquestionably to James Prescott Joule'.

Other sources for the study of Joule are: Lowery, 'The Joule Collection in the College of Technology, Manchester'; Rosenfeld, 'Joule's Scientific Outlook' and A. P. Hatton and L. Rosenfeld, 'An Analysis of Joule's Experiments on the Expansion of Air'.

29. Joule, *Scientific Papers* I, p. 267.

30. There are many questions which one asks at this point, and it will be well worth the effort to look into them. For instance it will be interesting to find out who was at the Oxford meeting, and what group interests were represented. Had Joule been from London or Cambridge, and not from Manchester, would then the chairman have asked him to cut it 'short'? What I imply here is supported by the well-known story that when Joule was asked in his later years what had he thought of the Royal Society's rejection of his first paper, Joule answered:

> I was not surprised, I could imagine those gentlemen in London sitting round a table and saying to each other, 'What could come out of a town where they dine in the middle of the day' [quoted by Rosenfeld, 'Joule's Scientific Outlook', p. 170, l. 33].

It is also interesting to note that the information reached the French almost immediately, and a French summary of Joule's paper appeared in the Comptes Rendus in August 1847, two months after the Oxford meeting. In the 1881 notes to his lectures Joule tells that the French paper was commissioned by Biot, Pouillet and Regnault. On his subsequent visit in Paris, Joule presented Regnault with his original iron vessel with its cooling paddle wheel.

31. Kelvin, *Mathematical and Physical Papers*, I, p. 102. It is well known that at the time of the Oxford meeting (1847) and when his first paper on this topic was published (1848) Kelvin had known about Carnot's theory only through Clapeyron's article (see Mendoza, *Reflections*, pp. 73–105). From his formulation of Carnot's main ideas it is clear that 'conservation of calorie' seems to him an essential part of Carnot's achievement. He writes:

> . . . it is by letting down of heat from a hot body to a cold body, through the medium of an engine (a steam engine or an air engine for an instance), that mechanical effect is to be obtained; and conversely he [Carnot] proves that the same amount of heat may, by the expenditure of an equal amount of labouring force, be *raised* from the cold to the hot body (the engine being in this case worked *backwards*) [Kelvin, *Mathematical Papers*, I, p... 103].

This made the establishment of an absolute scale possible for:

> The amount of mechanical effort to be obtained by the transmission of a given quantity of heat through the medium of any kind of engine in which the economy is perfect, will depend as Carnot demonstrates, not on the specific nature of the substance employed as the medium of transmission of heat in the engine but solely on the interval between the temperature of the two bodies between which the heat is transferred [ibid.].

A year later he received a copy of Carnot's original paper and he makes it clear that an absolute temperature scale can be derived from it if one accepts the caloric theory. 'The truth of this principle is considered as axiomatic by Carnot, who admits it is a foundation of his theory' (ibid., p. 115).

32. It is not germane to my argument how Kelvin reconciled the two theories. Let me remark only that he was very conscious about his task. He writes:

> The object of this paper is threefold:
> (1) To show what modifications of the conclusions arrived at by Carnot, and by others who have followed his peculiar mode of reasoning regarding the motive power of heat, must be made when the hypothesis of the dynamical theory, contrary as it is to Carnot's fundamental hypothesis, is adopted [ibid., I p. 176].

33. In an article in 1879 on 'thermodynamic motivity' Kelvin writes:

After having for some years felt with Professor Tait the want of a word 'to express the Availability for work of the heat in a given magazine, a term for that possession the waste of which is called Dissipation', I suggested three years ago the word *Motivity* to supply this want [ibid., I, p. 150].

34. Rankine, *Scientific Papers*, p. 235.
35. 'On the Centrifugal theory of Elasticity, as applied to Gases and Vapours', and 'On the Mechanical Action of Heat'.
36. Rankine, *Scientific Papers*, p. 235.
37. Kelvin, *Mathematical Papers*, I, p. 178.
38. In his 1852 notes to the 1851 paper Thomson added:

I take the opportunity of mentioning that I have only recently become acquainted with Helmholtz's admirable thesis on the principle of mechanical effect (Über die Erhaltung der Kraft, von Dr H. Helmholtz, Berlin, G. Reiner, 1847), having seen it for the first time on the 20th January of this year; and that I should have had occasion to refer to it on this, and on numerous other points of the dynamical theory of heat. . . .

Not only does Thomson not talk about the conservation of energy, but he sees in Helmholz's work merely another treatise on the dynamical theory of heat, and he even translates 'Über die Erhaltung der Kraft' as 'mechanical effect of heat' (Kelvin, *Mathematical Papers*, I, p. 183). He mentions Helmholtz again in a footnote on p. 244 and again only in this context.

39. Tait in his 1868 *Sketch of Thermodynamics* writes a historical introduction to the whole field. He talks in different chapters on the 'Dynamical Theory of Heat' (Chapter I) and on 'The Science of Energy' (Chapter II); yet when he repeats Thomson's two principles cited above he remarks:

. . . the second proposition (which regards the Transformation of heat as the first regards the Conservation of energy) . . .'.

This is one of the first sources to make this connection look self-evident.

40. Let me only mention the famous lectures by Judge Grove on the 'Correlation of Physical Forces' held at the London Institution in 1842, and which in view of the above described development reached immense popularity and was repeatedly reprinted (fifth edition in 1867). Also Mrs Somerville's well-known *On the Connexion of the Physical Sciences*. It appeared first in 1834, was reprinted eight times. The ninth revised edition of 1852 contains important additions on conservation of forces and on their mutual convertibility.

41. It is not accidental that I mention no name here. This simply reflects my belief that no problem is ever created by a purely scientific state of affairs.

42. Y. Elkana, 'Helmholtz's "Kraft": an Illustration to Concepts in Flux'.

43. As this point occupies a major part of my essay on Helmholtz, I shall only mention here briefly the argument:

The nineteenth century has inherited two basically different traditions in mechanics: Newtonian-vectorial mechanics, with its emphasis on forces, and the Leibniz-Euler-Lagrange formulation of analytical mechanics with its emphasis on the scalar quantities of the *vis viva* and the potential function. The major concern of vectorial mechanics as formulated by the Newtonians was to measure the action of a force by its momentum—an approach that originated with Descartes; Descarte's momentum, however, is undoubtedly a scalar quantity, and as such serves as foundation to both traditions. The basic concepts in Newtonian mechanics were space, time, mass, and force. The drawbacks of these formulations were that for cases where constraints occurred the treatment became rather tedious; besides, the action-reaction law does not embrace all cases—it proves to be sufficient actually only for dynamics of rigid bodies. Its great advantage was that forces which are not derivable from a work function, i.e. are not conservative but of frictional nature, and cannot be dealt with by the mechanical energy conservation principle, can be treated easily by Newtonian mechanics. Needless to say, the apparent loss of *vis viva* in frictional processes troubled much less those physicists who could still succeed in solving the basic dynamical problem which they set to themselves. On the other hand the basic concepts in the Euler-Lagrange procedure were space, time, mass and energy. This procedure is applicable only to forces which are conseravtive, that is to say, depend only on position and not on time or velocity. Here the conservation of mechanical energy holds; these are conservative systems.

44. The emphasis is on a *mathematician's certainty*, because dimensional analysis as a decisive factor in pure physics was in its infancy.

45. Faraday's 'sense' of force is clear in the following statement:

What I mean by the word [force] is the source or sources of all possible actions of the particles or materials of the universe.

On that topic more can be found in my paper on Helmholtz.

46. Helmholtz: 'Über den Stoffverbrauch bei der Muskelaktion'; 'Über das Wesen der Faulniss und Gährung'; 'Breichte über die Theorie der physiologischen Wärmeerscheinungen'. All in his *Wissenschaftliche Abhandlungen*.

47. Helmholtz, 'On the Conservation of Force', translated by J. Tyndall in *Taylor's Scientific Memoirs*, I (1852).

48. ibid., p. 131.

49. Joule, *Scientific Papers*, I, p. 202.

50. Rankine, *Scientific Papers*, p. 203.

51. 'The law of the Conservation of Energy is already known—viz. that the sum of all of the energies of the universe, actual and potential, is unchangeable.' ibid.

52. ibid., p. 201.

53. ibid., p. 209.

Bibliography

Adickes, E., *Kant als Naturforscher*, vols. i & ii (Berlin, 1925).

Agassi, J., 'Toward a Historiography of Science', vol. 2 to *History and Theory* (1963).

—— 'The Nature of Scientific Problems and their Roots in Metaphysics' in M. Bunge (Ed.), *The Critical Approach to Science and Philosophy* (Glencoe, 1964), pp. 189–211.

—— *Faraday as a Natural Philosopher* (Chicago, 1971).

d'Alembert, J. le R., 'On Force', item in the *Grande Encyclopédie*.

——*Traité de Dynamique*, Ed. Gauthiers-Villars (1921).

Allamond, Jean Nic. Seb. (Ed.), *Œuvres Philosophiques de Mr. G. G.'s Gravesande* (Amsterdam, 1774).

Andrade, E. N. da C., 'Two historical notes', *Nature* (9 March 1935).

Appelyard, R., *A Tribute to Faraday* (London, 1931).

Aristotle, *Physica*, vol. ii of W. D. Ross (Ed.), *The Works of Aristotle* (Oxford, (1953).

Bacon, Sir Francis, *The works of Francis Bacon*, R. L. Ellis, J. Spedding and D. D. Heath (Eds.) (London, 1859; reprinted Friedrich Fromann Verlag, Stuttgart 1965).

Bairoch, P., 'Original Characteristics and Consequences of the Industrial Revolution', *Diogenes*, **54**, p. 47.

Barber, B., and Hirsch, W. (Eds.), *The Sociology of Science* (New York, 1962).

Ben-David, J., Review of Roger Hahn: 'The Anatomy of a Scientific Institution: The Paris Academy of Sciences, 1666–1803' in *Minerva*.

—— *The Scientist's Role in Society* (Prentice Hall, 1971).

Bernal, J., *The Social Function of Science* (M.I.T. Press, 1967).

Berzelius, J. J., *Lehrbuch der Chemie* (Wohler, 4 vols., Dresden, 1825–31).

Blackwell, R. J., Spath, R. J., and Thirkell, W. E., 'Commentary on Aristotle's Physics by St Thomas Aquinas' (Yale, 1963).

Bohm, D., 'Space, Time and the Quantum Theory', 1966 (lecture notes).

Bornsted, Y. N., *Energetics* (New York, 1952).

Brett, R. L. (Ed.), *S. T. Coleridge* (G. Bell and Sons, 1971).

Brooke, C., 'Force and Energy', *Nature* (1872), p. 122.

Brown, S. C., 'Count Rumford's Concept of Heat', *Amer. J. Phys.* **20** (1952).

—— 'The Caloric Theory of Heat', *Amer. J. Phys.* **18** (1950), pp. 307–73.

—— *Count Rumford, Physicist Extraordinary* (Anchor Books, 1952).

——*Benjamin Thomson, Count Rumford* (Pergamon Press, 1966).

Bulwer-Lytton, E. G., *England and the English* (London, 1833).

Cardwell, D. S. L., *From Watt to Clausius* (Heinemann, London, 1971).

Carnot, H., 'Lettre Adressée à l'Académie des Sciences' (Paris, 1878).

Carnot, S., *Biographie et Manuscrit* (Gauthier-Villars et Cie., Paris, 1927).

Cleghorn, R. A., *Disputatio Physica Inauguratis, Theoriam Ignis Competentur* (Edinburgh, 1773).

Coburn, K. (Ed.), *The Philosophical Lectures of S. T. Coleridge* (London, 1949).

Cohen, I. B. (Ed.), *Isaac Newton's Papers and Letters on Natural Philosophy* (Cambridge, 1958).

—— The Wiles Lectures held at the Queen's University, Belfast, in 1966, to be published.

—— *Introduction to Newton's Principia* (Cambridge, 1971).

Colding, L., 'On the History of the Principle of the Conservation of Energy', *Lond. Phil. J.* (1864).

Cranefield, P. F., 'The Organic Physics of 1847 and Biophysics Today', *J. Hist. Med.* **xii** (1957).

—— 'The Philosophical and Cultural Interests of the Biophysical Movement of 1847', *J. Hist. Med.* **xxi** (1966).

Dahl, P. F., 'Ludvig Colding and the Conservation of Energy', *Centaurus* **8** (1963).

Davy, J., *Memoirs of the Life of Sir Humphry Davy, Bart.*, 2 vols. (London, 1836).

—— *The Collected Works of Sir Humphry Davy* (London, 1839–40) in 9 vols; vol. i is the biography.

—— 'Fragmentary Remains, Literary and Scientific of Sir Humphry Davy, Bart., With a Sketch of His Life and Selections from His Correspondence' (London, 1858).

De Miran, 'Dissertation sur l'Estimation et la Mesure des Forces Matrices des Corps' (1728).

Dijksterhuis, E. J., *The Mechanization of the World Picture*, (Oxford, 1970).

—— *Simon Stevin Science in the Netherlands around 1600* (The Hague, 1970).

Disraeli, B., *Coningsby* (London, 1844).

Dricks, H., *Perpetual Motion in the 17th and 18th Centuries*, (London, 1861).

Du Bois-Reymond, E., *Reden*, vols. i and ii (Verlag von Veit and Co., Leipzig, 1912).

Dugas, R., *History of Mechanics* (Geneva, 1955).

Dühring, E., *Kritische Geschichte der allgemeinen Principien der Mechanik* (Berlin, 1872).

Dulong, P. L., and Petit, A. T., *Ann. Chem. Phys.* **7** (1818).

Einstein, A., *Herbert Spencer Lecture* (Clarendon Press, Oxford, 1933).

Elkana, Y., 'Ludwig Boltzmann: "On the Development of the Methods of Theoretical Physics in Recent Times" ', *The Philosophical Forum*, **I** (1968), pp. 94–120.

—— 'Science, Philosophy of Science and Science Teaching', *Educ. Phil. Theory*, **2** (1970), pp. 15–35.

—— 'Helmholtz's "Kraft": An Illustration of Concepts in Flux', *Hist. Stud. Phys. Sci.*, **2** (1970), pp. 263–298.

—— 'The Conservation of Energy: A Case of Simultaneous Discovery?', *Archives Internationales d'Histoire des Sciences*, 23–e année, **90–I** (1970), pp. 31–60.

—— 'David M. Knight's "The Concept of the Atom" ', *Science*, **163** (1969), pp. 378–79.

—— 'Newtonianism in the Eighteenth Century', review essay, *Brit. J. Phil. Sci.*, **22** (1971), pp. 297–306.

—— 'The Problem of Knowledge', *Studium Generale*, **24** (1971), pp. 1426–1435.

—— 'The Historical Roots of Modern Physics', four lectures delivered at the International School of Physics 'Enrico Fermi', 'Topics in the History of 20th Century Physics', Varenna, August 1972 (to appear in the proceedings of the Varenna Meeting, Academic Press, ed. Ch. Wedner).

——'Scientific and Metaphysical Problems: Euler and Kant', in *Boston Stud. Phil. Sci.*, **xiv** (1974), pp. 277–305.

—— 'Boltzmann's Scientific Research Programme and its Alternatives' in Y. Elkana (Ed.), *The Interaction between Science and Philosophy* (Humanities Press, 1974)

Epstein, S. T., *Textbook of Thermodynamics* (New York, 1934).

Euler, L., 'Mechanica sive Motus Scientia Analytica Exposita', Ed. P. Staeckel, in *Opera Omnia series Secunda*.

Faraday, M., 'On the Conservation of Force', *Proc. Roy. Soc. Lond.* (1857), republished in the *Phil. Mag.* (1857).

—— *On the Various Forces of Nature* (New York, The Viking Press, 1960).

Fullmer, J. Z., 'On the Poetry of Sir Humphry Davy', *Chymia* **4** (1958).

Gillispie, C. C., *Lazare Carnot, Savant* (Princeton, 1971).

Goodfield, J. G., *The Growth of Scientific Physiology* (London, 1960).

Greene, G., 'Applications of Mathematical Analysis to Electricity and Magnetism', as quoted by the *O.E.D.* under item 'Potential'.

Grove, W. R., *The Correlation of Physical Forces* (London, 1855).

Guerlac, H., 'J. Black and Fixed Air, etc.' *Isis*, **48** (1957).

—— *Lavoisier—The Crucial Year* (Cornell, 1961).

Haene, R. D', 'La notion scientifique de l'énergie, son origine et ses limitations', *Revue de Metaphysique et Morale* (1967), p. 35.

Hahn, R., *The Anatomy of a Scientific Institution, The Paris Academy of Sciences, 1666–1803* (California, 1971).

Haller, A. von, *First Lines of Physiology* (Edinburgh, 1801).

Hankins, T. L., *Jean D'Alembert: Science and the Enlightenment* (Oxford, 1970).

——'Eighteenth-Century Attempts to Resolve the Vis-Viva Controversy', *Isis*, **lxi** (1965).

Hare, R., *Compendium of the Cause of Chemical Instruction* (1828).

Hartley, Sir H., *Humphry Davy* (London, 1966).

Hatten, A. P. and Rosenfield, L., 'An Analysis of Joule's Experiments on the Expansion of Air', *Centaurus*, **4** (1956), pp. 311–18.

Hell, B., *J. Robert Mayer und das Gesetz von der Erhaltung der Energie* (Stuttgart, 1925).

Helmholtz, H., von, 'Ueber den Stoffverbrauch in der Muskelaktion', *Mueller's Archiv* (1845), or in vol. 2 of the *Wissenschaftliche Abhandlungen*.

——'Bericht ueber die Theorie der Physiologischen Waermeerscheinungen' *Fortschritte der Physik* for 1845, appeared in 1847, or in *Wissenschaftliche Abhandlungen*, vol. 1.

—— Notes added to the original 1847 paper 'Über die Erhaltung der Kraft' in the 1881 ed. of his *Wissenschaftliche Abhandlungen.*

—— John Tyndall's translation, *Taylor's Scientific Memoirs* (1854).

—— Preface to the German ed. of J. Tyndall's *Faraday as a Discoverer*, published in *Nature* (1870), p. 51.

—— Preface to the German ed. of W. Thomson and P. G. Tait's *Natural Philosophy*, trans. Professor C. Brown and published in *Nature* (1874), p. 149.

—— 'Faraday Lecture', published in *Nature* (1881), p. 535.

—— *Wissenschaftliche Abhandlungen*, 3 vols. (Leipzig, 1882).

—— Arthur Koenig et. al. (Eds.), *Vorlesungen ueber Theoretische Physik von H. von Helmholtz*, 6 vols. (Leipzig, 1897).

—— *Vortraege und Reden*, 2 vols. (Braunschweig, 1903).

Herivel, J., 'Prerequisites for Creativity in Theoretical Physics', *Scientia* **60** (1966).

Herivel, Prof. J., 'The Background to Newton's *Principia*' (Oxford, 1965).

Hertz, H., *Gesammelte Werke*, vol. i, *Schriften Vermischten Inhalts* (Johann Ambrosius Barth, Leipzig, 1895).

Hessen, B., 'On the Social and Economic Roots of Newton's *Principia*' in *Science at the Cross Road* (Cass Reprints, 1971).

Hiebert, E. N., *The Historical Roots of the Principle of Conservation of Energy* (Madison, 1962).

——— Commentary to Kuhn's paper in *Critical Problems in the History of Science*, M. Clagett (Ed.), pp. 391–8.

———'The Uses and Abuses of Thermodynamics in Religion', *Daedalus Fall* (1966), p. 1046.

Histoire de l'Académie Royale des Sciences, to. 4 (Paris, 1775).

Jones, B., *The Royal Institution* (London, 1971).

Joule, J. P., *Joule's Scientific Papers*, vols. i, ii (Dawson Reprint).

Jørgensen, B. S., 'Berzelius und die Lebenskraft', *Centaurus* **10** (1964).

Kant, I., *Critique of Pure Reason*, trans. N. V. Smith (New York, 1965).

Katchalsky, A., and Curran, P. F., *Nonequilibrium Thermodynamics in Biophysics* (Harvard U.P., 1965).

Keeton, M. T., 'Some Ambiguities in the Theory of the Conservation of Energy', *Phil. Sci.* (1940).

Kelvin, Lord W. T., 'On a Universal Tendency in Nature to the Dissipation of Mechanical Energy', *Trans. Roy. Soc. Edinburgh* **20** (1853).

——— *Mathematical and Physical Papers*, i.

——— *Collected Papers* (Cambridge, 1911).

Klicstein, H., 'C. Caldwell and the Controversy in America over Liebig's "Animal Chemistry",' *Chymia*.

Knott, C. G., *The Scientific Work of P. G. Tait* (Cambridge, 1911).

F. O. Koenig, 'On the History of Science and of the Second Law of Thermodynamics,' in H. M. Evans (ed.), *Men and Moments in the History of Science* (University of Washington Press, 1959).

Koenigsberger, L., 'The Investigations of Hermann von Helmholtz on the Fundamental Principles of Mathematics and Mechanics', *Ann. Rep. Smiths. Inst.* (1890).

——— *H. von Helmholtz*, 3 vols. (Braunschweig, 1901).

Krebs, G., *Die Erhaltung der Energie als Grundlage der Neueren Physik* (Muenchen, 1877).

Kuhn, T. S., 'Energy Conservation as an Example of Simultaneous Discovery', published in M. Clagett (Ed.), *Critical Problems in the History of Science* (Wisconsin, 1955), pp. 321–56.

——— 'Carnot's Version of "Carnot's Cycle",' *Amer. J. Phys.* **23** (1955) p. 31.

Lagrange, J. P., *Mécanique Analytique* (Paris, 1788).

Lakatos, I., 'History of Science and its Rational Reconstructions' in Y. Elkana (Ed.), *The Interaction between Science and Philosophy* (Humanities Press, 1974).

——— 'Falsification and the Methodology of Scientific Research Programmes' in I. Lakatos and A. Musgrave (Eds.), *Criticism and the Growth of Knowledge* (Cambridge, 1970).

La Mer, V. K., 'Some Current Misinterpretations of N. L. Sadi Carnot's Memoir and Cycle', *Amer. J. Phys.* **21** (1953); **22** (1954), p. 20; **23** (1955), p.95.

Lanczos, C., *The Variational Principle in Mechanics*, (Toronto, 1954).

Lasswitz, K., *Geschichte der Atomistik vom Mittelalter bis Newton*, vols. i–iv (Georg Olms, 1963).

Lavoisier, A., *Traité Elémentaire de Chimie* (Paris, 1789).

—— *Essays, Physical and Chemical*, trans. T. Henny (reprinted by F. Cass, 1970).

Lavoisier, A., and Laplace,, P. S., *Mémoire sur la Chaleur*, Gauthier-Villars (Eds.)

Lilley, S., 'Attitudes to the nature of heat about the beginning of the nineteenth century', *Arch. Intern. d'Histoire des Sciences* **27** (1948), p. 630.

—— 'Cause and Effect in the History of Science', *Centaurus* **3** (1953), p. 53.

Lipman, T. O., 'The Response to Liebig's Vitalism', *Bull. Hist. Med.* **xl** (1966).

Lodge, Sir O., 'On the Seat of the Electromotive Forces in the Voltaic Cell', *The London, Edinburgh and Dublin Philosophical Magazine and Journal of Science*, 5th ser. March 1885).

—— *Energy* (London, 1929).

Loemker, (Ed.), *Leibniz's Philosophical Papers and Letters* ,2 vols. (Chicago, 1955).

Lowery, J., 'The Joule Collection in the College of Technology, Manchester', I. *J. Sci. Instrum.*, **vii** (1930), p. 369; II, **viii** (1931), p. 1.

Mach, E. *History and Root of the Principle of Conservation of Energy*, trans. P. E. B. Jourdain (The Open Court Publishing Co., Illinois, 1911).

—— *Popular Scientific Lectures* (Open Court Publishing Co, Illinois, 1943).

—— *The Science of Mechanics: A Critical and Historical Account of its Development* (Open Court Publishing Co, Illinois, 1960).

Mackenna, S., *Plotinus' Psychic and Physical Treatises* (London, 1921).

Martin, T., *The Royal Institution* (London, 1941).

Maxwell, J. C., 'Scientific Worthies: Hermann von Helmholtz', *Nature*, **xvi** (1877).

—— *Matter and Motion* (Dover, 1965).

Mayer, J. R., 'Bemerkungen ueber die Kraefte der unbelebten Natur', *Ann. Chem. Pharm.* (1842). Reprinted in *Raum und Zeit*, Ed. E. Wildhagen (Deutsche Buch-Gemeinschaft, Berlin).

Mckendrick, J. G., *Hermann Ludwig Ferdinand von Helmholtz* (New York, 1899.

McKie, D., *Nature*, **153** (1944).

—— *Lavoisier* (London, 1952).

Mendelsohn, E., *Heat and Life* (Harvard, 1964).

—— 'The Biological Sciences in the Nineteenth Century: Some Problems and Sources', *History of Science* **3** (1964).

—— 'The Continuing Scientific Revolution', lecture in the Harvard Summer School series.

Mendenhall, T. C., 'Helmholtz', *Ann. Rep. Smiths. Inst.* (1895).

Mendoza, E. (Ed.), *Reflections on the Motive Power of Fire by Carnot and other papers on the Second Law of Thermodynamics by E. Clapeyron and R. Clausius* (Dover, 1966).

Merton, R. K., 'Singletons and multiples in scientific discovery: a chapter in the sociology of science', *Proc. Amer. Phil. Soc.*, **105** (Oct. 1961), pp. 420–86.

—— 'Resistance to the Systematic Study of Multiple Discoveries in Science', *European Journal of Sociology*, **4** (1963).

Merz, J. T., *A History of European Thought in the Nineteenth Century*, 4 vols (Dover, 1965).

Metzger, H., *Les Doctrines Chimiques en France* (Paris, 1969).

Meyerson, E., *Identity and Reality* (Dover, 1962).

—— 'Y-a-t-il un rhythme dans le progrès intellectuel?', in *Essais* (Paris, 1936)

Moore, J., 'The Conservation of Energy not a Fact, but a Heresy of Science', *Nature* (1872), p. 180.

—— 'The Heresies of Science', *London Quarterly Review* (1872).

Møller, C., 'The Concept of Mass and Energy in the General Theory of Relativity', DKNVS Forhandlinger, **31** (1958).

Nicolson, J., 'The Conservation of Force', *Nature* (1871), p. 47.

Ørsted, H. C., *The Soul in Nature* (H. G. Bahn, London, 1852) (reprinted by Dawson of Pall Mall, London, 1966).

Ogburn, W. F., and Thomas, D. S., 'Are Inventions Inevitable?', *Political Science Quarterly* 37 (1922), p. 83.

Ostwald, G., *Wilhelm Ostwald_1 Mein Vater* (Berliner Union, Stuttgart, 1953).

Ostwald, W., *Vorlesungen ueber Naturphilosophie* (Leipzig, 1902).

—— *Grosse Maenner* (Leipzig, 1903).

—— *Die Energie* (Leipzig, 1908).

—— *Monism as the Goal of Civilization* (1912).

Paris, J. A., *The Life of Sir Humphry Davy*, vols i & ii. (Henry Colburn and Richard Bently, 1831).

Partington, G. R., *A Short History of Chemistry* (London, 1957).

—— 'J. Black's Lectures on the Elements of Chemistry', *Chymia* 5 (1959).

Patterson, L. D., 'Robert Hooke and the Conservation of Energy', *Isis* 38 (1948).

Pauli, W., 'Phenomaen and Wirklichkeit' in the collection *Aufsaetze und Vortraege ueber Physik und Erkenntnistheorie* (Braunschweig, 1961).

Plaass, P., *Kant's Theorie der Naturwissenschaften* (Goettingen, 1965).

Planck, M., *Das Prinzip der Erhaltung der Energie*, 4th ed. (Berlin, 1921).

Pledge, H. T., *Science since 1500* (Harper, 1959).

Poincaré, H., 'Energy and Thermodynamics' in *Science and Hypothesis* (Dover, 1952).

Popper, Sir K., 'The Nature of Philosophical Problems and their Roots in Science' in *Conjectures and Refutations* (London, 1963).

Rankine, W. J. M., *Miscellaneous Scientific Papers* (London, 1880).

—— 'On the Mechanical Action of Heat', *Trans. Royal Soc. of Edinburgh* xx (1850).

—— 'On the Centrifugal Theory of Elasticity, as applied to Gases and Vapors', *Phil. Mag.* (Dec. 1851).

Ravetz, J., 'The Evolution of Science and its History', *Acta historiae rerum naturalium nec non technicarum* (Prague, 1967).

Reynolds, O., 'Memoir of J. P. Joule', *Mem. Proc. Manchester Lit. and Phil. Soc.* (4), 6 (1892), p. 179.

Riese, W., and Arrington, G. E., 'The History of Johannes Mueller's Doctrine of the Specific Energies of the Senses: Original and Later Version', *Bull. Hist. Med.* 37 (1964).

Rohault, J., *Traité de Physique* (Paris, 1671).

Rosenfeld, L., 'Joule's Scientific Outlook', *Bull. Br. Soc. Hist. Sci.* i (1952), pp. 169–76.

Ruecker, A. W., 'F.R.S. "Helmholtz",' *Ann. Rep. Smiths, Inst.* (1894).

Rumford, Count B., *Mémoire sur la Chaleur* (Paris, 1804).

—— 'Experimental Inquiry Concerning the Source of the Heat which is Excited by Friction', *Phil. Trans.* lxxxvii (1798), p. 80.

Sabra, A. I., *Theories of Light from Descartes to Newton* (Oldbourne, London, 1967)

Sambursky, S., *The Physical World of Late Antiquity* (London, 1962).

Sarton, G., 'The Discovery of the Law of Conservation of Energy', *Isis*, xiii (1929), p. 18.

Schlick, M., *Philosophy of Nature* (Philosophical Library, New York, 1949).

Schofield, P., 'The Lunar Society of Birmingham' (Oxford, 1953).

Scott, W. L., 'The Significance of "Hard Bodies" in the History of Scientific Thought', *Isis* 50 (1959), pp. 199–210.

Seguin, M. A., 'Mémoire sur les Causes et sur les Effets de la Chaleur de la Lumière et de l'Electricité' (Paris, 1865).

Siegfried, R., 'The Chemical Philosophy of Humphry Davy', *Chymia* 5 (1959).

Simmel, G., *Kant* (Leipzig, 1904).

Somerville, Mrs., *On the Connexion of the Physical Sciences* (1834; 9th revised ed., 1852).

Stallo, J. B., 'The Concepts and Theories of Modern Physics', Ed. P. W. Bridgman (Harvard).

Stewart, B., *The Conservation of Energy* (London, 1874).

Tait, P. G., *Sketch of Thermodynamics*, (1868).

Thackray, A., *Atoms and Powers* (Cambridge, Mass., 1970).

Theobald, D. W., *The Concept of Energy* (London, 1966).

Thomson, W., and Tait, P. G., *Lectures on Natural Philosophy* (Cambridge, 1879).

Thorpe, Sir E. (Ed.), *The Unpublished Scientific Papers of Henry Cavendish* (Cambridge, 1921).

Tisza, L., 'The Conceptual Structure of Physics', *Rer. Mod. Phys.* 35 (1963), p. 343.

—— *Generalized Thermodynamics* (M.I.T. Press, 1966).

Truesdell, C., 'A Program toward Rediscovering the Rational Mechanics of the Age of Reason', *Arch. Hist. Exact Sciences* (1955).

—— *Essays in the History of Mechanics* (Berlin, 1968).

Tyndall, J., *Heat as a Mode of Action* (London, 1804).

—— *Heat Considered as a Mode of Motion* (New York, 1863).

Walden, P., *Naturwissenschaften*, 16 (1928).

Watanabe, M., 'Count Rumford's First Exposition of the Dynamical Aspect of Heat', *Isis*, 50 (1959).

Watson, E. C., 'Joule: Only General Exposition of the Principle of Conservation of Energy', *Amer. J. Phys.* 15 (1947), p. 383.

Weisheiple, J. A., *The Development of Physical Theory in the Middle Ages* (New York, 1959).

Westfall, R. S., *Force in Newton's Physics* (New York, 1971).

Wiener, P. P. (Ed.), *Selections from Leibniz* (Scribners, 1951).

Williams, L. P., *Michael Faraday* (Basic Books, 1966).

Williams, L. P. (Ed.), *The Selected Correspondence of Michael Faraday*, 2 vols. (Cambridge, 1971).

Wood, A., *Joule and the Study of Energy* (London, 1934).

—— *Thomas Young*, Ed. Frank Oldham (Oxford, 1954).

Youmans, *The Correlation and Conservation of Forces* (Appleton N.Y.C., 1865).

Young, T., *Lectures on Natural Philosophy* (London, 1807).

Index